BUILDING ECOLOGY
FIRST PRINCIPLES FOR A SUSTAINABLE BUILT ENVIRONMENT

BUILDING ECOLOGY

FIRST PRINCIPLES FOR A SUSTAINABLE BUILT ENVIRONMENT

PETER GRAHAM

Lecturer – Sustainable Construction
Department of Building and Construction Economics
RMIT University
Melbourne, Australia

Blackwell
Science

© 2003 by Blackwell Science Ltd,
a Blackwell Publishing Company
Editorial Offices:
Osney Mead, Oxford OX2 0EL, UK
Tel: +44 (0)1865 206206
Blackwell Science, Inc., 350 Main Street,
Malden, MA 02148-5018, USA
Tel: +1 781 388 8250
Iowa State Press, a Blackwell Publishing
Company, 2121 State Avenue, Ames, Iowa
50014-8300, USA
Tel: +1 515 292 0140
Blackwell Science Asia Pty Ltd, 550
Swanston Street, Carlton South,
Victoria 3053, Australia
Tel: +61 (0)3 9347 0300
Blackwell Wissenschafts Verlag,
Kurfürstendamm 57, 10707 Berlin, Germany
Tel: +49 (0)30 32 79 060

Line drawings by Anita Lincolne-Lomax,
Lincolne-Lomax Design, Melbourne,
Australia

First published 2003 by Blackwell Science Ltd

Library of Congress
Cataloging-in-Publication Data
is available

ISBN 0-632-06413-7

A catalogue record for this title is available
from the British Library

Set in 10.5/12.5 pt Palatino
by Sparks Computer Solutions Ltd, Oxford
http://www.sparks.co.uk
Printed and bound in Great Britain by
MPG Books Ltd, Bodmin, Cornwall

For further information on
Blackwell Science, visit our website:
www.blackwell-science.com

Dedicated to my Nan

Phyllis Moulton

CONTENTS

Contents

Contents

ACKNOWLEDGEMENTS

This book has not come from me, but through me. I have been facilitator of the ideas presented here, more developer than creator, and as such have many to thank for this final outcome. I would not have been able to complete this without the love and support of my parents – Dr Kay Moulton and Dr John Graham and my sister Sarah. Having a family as eminent and willing to read and comment on the many drafts is a wonderful thing. I must especially thank my mother for letting me stay in the spare room and for feeding me wonderfully during the last two weeks of writing. For her passion for life, her love and inspiration I would like to thank Carly Davenport, whose confidence in my capacity to write a book has kept me going. In my darker moments you reminded me of our love for life, and the urgency of our campaign to make the world a better place for all children. You are beautiful.

Bill Lawson, from the University of New South Wales – thank you for your excellent comments on my draft manuscript. Without your body of work, this book would have been very hard to complete. Thank you for asking me whether I use 'nature' in the big 'N' sense, as designed by some 'transcendental force', 'first cause' or 'God'. You will notice that I have opted for the small 'n' as in 'evolving systems'.

As a surfer I certainly feel big 'N' nature and this fuels my passion for life. In writing this book the little 'n' comes out because I don't want to condition people's understanding with my perspective alone. I am sure that everyone has some perception of the presence or otherwise of a divine energy. The imperative is for us all to agree on what forms of development will keep us alive so we can enjoy our exploration of what nature means to us. The current state of the world exemplifies what happens when people's attachment to ideas are given priority over developing understandings about what we all share.

Building with nature means understanding how to work with the physical laws that govern life. Any surfer will tell you what it is like. Surfers learn the art of going with the flow – not aimlessly following a whim, but directing their action to find the location and the moment in which all of the energies present in the wave work together. I would like to thank my friends in Urban Ecology and the Surfrider Foundation for all of our work together, our conversation and inspiration. Bill and Heather Pemberton deserve a special mention for reading, then staying up very late debating thermodynamics with me. Thanks also to all of the people that contributed ideas and case studies.

A big thank you also to Mary Tomsic for her exceptional research assistance and compilation of many of the tables. *Building Ecology* would not have been completed within deadline had it not been for your calm, 'can do' attitude.

My students at RMIT in Australia and SIM in Singapore used the text and tried out the activities. Their feedback provided insights into approaches to teaching and learning that have helped shape this book. I must also thank my colleagues at the Department of Building and Construction Economics at RMIT for their support.

This book is a consequence of the relationships created between people. It is a manifestation of an emerging system emblematic of our time – an empowered network. A saying common in the twentieth century was 'it is not what you know – but who you know'. The empowered network is more than just knowing people. It is about helping people. A saying for the twenty-first century that might define an empowered network is 'it is not just who you know – but how you help each other'. I would not have been able to write *Building Ecology* without my empowered network. Thank you.

Peter Graham
February 2002

ABOUT THIS BOOK

This book is intended for students and building practitioners who are new to the field of ecologically sustainable building. It is designed to provide grounding in the general issues and in the fundamental laws and principles of ecological sustainability. It is written for all people involved in shaping our built environment, not for any one particular vocation.

I have written this book because I want to provide a way of helping people understand their intimate interdependency with nature, and be able to make decisions that sustain and can be sustained by nature. I have not set out to tell people what to do, or what to think about building or the environment. I am interested in engaging how we think about building and the environment and the models we think with.

I have set out to provide a more holistic, systems-based way of thinking about ecologically sustainable building that can help people develop an understanding of the relationships between natural systems and built environments and between decisions and intended outcomes. The emphasis therefore is not on providing comprehensive descriptions of environmental impacts of building materials, nor extensive building design critiques.

Building Ecology is intended to help people develop an initial level of ecological literacy and environmental awareness, an approach to thinking about building that can become a platform for life-long learning in this field.

The book is structured in three sections. The first part deals with how our buildings and our minds are connected with natural systems. Its purpose is to increase understanding of how and why building affects nature. It introduces the concepts of life-cycle thinking and urban metabolism as ways of understanding relationships between building and nature. The impacts of our current practice on biogeochemical cycles and ecosystems are then described. We discover that our building activity is causing changes in natural systems that are having a negative impact on us. We need to change…but how?

Part II deals with that question of change. It describes the fundamental principles of ecological sustainability derived from thermodynamic laws, and contains observations of the way we are changing nature, and in turn the pressures nature places on us to change. At the end of section two we understand what the properties of ecologically sustainable building are.

The purpose of Part III is to take a look again at the big picture, but this time from an ecologically sustainable perspective. This section distils a set of laws and principles for ecologically sustainable building. These principles are presented in a framework that makes sense of their individual and collective purpose. We then look at how a built environment might progress if these principles were applied. A detailed case study of a school in Sweden is also provided in this section to show the sort of building that results from a holistic approach to thinking about ecological sustainability.

As general rather than detailed knowledge is presented, suggestions for supplementary and complementary reading and information resources are provided at the end of each section, organised into topic areas. Reflective learning activities are also provided at the end of Parts I and II as tools for assisting deeper thinking on issues raised.

Many terms used require definition or a level of explanation for which space in the text does not allow. A glossary of terms is provided for this purpose.

I hope you enjoy *Building Ecology.*

1 INTRODUCTION

People and buildings are as intrinsically infused with nature as bees and their hives are with honey. Unfortunately the effects of our building industry are far less sweet. Unlike bees, whose produce is nourishing, the process of constructing and operating buildings often creates environmental outcomes that are not. Due to the non-ecological structure of the building industry and the historical lack of environmental awareness of many building professionals, the way buildings, built environments and the process of building have been created, has played a major role in the decline in Earth's ecological health.

Ecosystems are life-supporting systems and building that can be sustained by ecosystems rather than damage them is urgently required. The intent of this book therefore, is to provide a way of extending the understanding of building from mere regard for the structure and material to include all of the natural systems with which all building is interdependent. It also extends the concept of the building professional from one who helps make a structure, to one who builds a system of relationships between ecosystems and human systems. In essence this book is designed as an introduction to a systems approach to thinking about building for people new to the field of ecologically sustainable building.

Modern science is providing a far more detailed perspective of the way ecosystems function. This new knowledge is now beginning to be used by building professionals and the building industry in order to protect the life-supporting goods and services of ecosystems, and provide the basis for sustained life opportunity on Earth. *Building Ecology* presents approaches to gaining knowledge about the environmental influences of building in a holistic framework that helps students of building professions and practising professionals to:

- understand the interdependencies of buildings and nature;
- understand how building affects nature;

- understand what is and is not sustainable;
- understand how building can work with nature.

A hive of understanding

Imagine all of the existing knowledge on ecologically sustainable building as cells of honeycomb in a beehive, and all knowledge of economics and all other fields of human enquiry and understanding as being brought together in a honeycomb structure. The knowledge within each of these cells is like the pollen that busy bees go out to collect, while the processed and applied knowledge becomes understanding – and that is the honey.

Experts of many disciplines are busy at work in and around their cells collecting and processing knowledge in order to create understanding which, like honey, nourishes us. In building we must bring together many types of knowledge, apply many types of understanding. As well as thinking within a cell of specialised knowledge, we have to think *outside* the cell in order to solve the complex problems posed by creating a built environment in a world of decreasing ecological health and increasing human demands.

How do we learn to select, and then fit together, cells containing the knowledge we need, in a way that allows us to appreciate the whole system that we will affect when we build? We need a model that can allow us to bring together the knowledge contained in discrete disciplinary cells, and which allows the honey of combined understanding to flow freely.

A honeycomb in a beehive has a unifying hexagonal pattern, which is common to all cells. Too often environmental courses at universities are offered as separate subjects to those dealing with the perceived 'world' of financial and economic 'reality'. Cells of knowledge concerning such issues as the environmental impact of materials, energy efficiency, sustainable construction, environmental design and management, cost estimating and quantity surveying, to name a few, although essential for holistic understanding of ecologically sustainable building, are sometimes as fragmented in curricula as a beehive might be after a bear's breakfast.

The media too, frequently sets issues of environmental health in opposition to economic and social welfare, reinforcing the perception of a chasm between such things as 'jobs and forests' or the 'cost of energy-efficient building and greenhouse gas reduction'. 'Building bridges' of compromise is then required where the outcomes are not optimal for anyone.

2

Introduction

All of us are affected by the fragmented or adversarial way issues are presented to us. Many have accepted, perhaps unconsciously, what Gregory Bateson[1] refers to as the 'obsolete' Cartesian dualism of mind and matter. Another 'obsolete' idea that regularly enters the classroom is that economics simply refers to the management of financial transaction and that it is an opposing idea to that of ecological sustainability, further fragmenting our worldviews. So in learning ecologically sustainable building we end up being required to integrate discrete cells of knowledge using our own fragmented mental models of the way we think the world works. Unfortunately, when it comes to sustaining the life-supporting processes of our planet there is no room for compromise.

If we are going to be successful in creating ecologically sustainable building then we need to deal with the issue of how we have been constructed to see the world. We need to be given thinking tools that we can use to first develop a holistic rather than dualistic worldview and then use to reintegrate existing knowledge. As honeycomb has its hexagon, our knowledge and understanding of ecologically sustainable building needs a unifying pattern also.

In nature there exists a common thread of influences that, considered together, provide a model for synthesising existing knowledge so that we can understand it in the context of the broader ecological systems. This common thread allows us to understand our effects on the natural systems within which our economies sit, not as separate to economic concerns, but as integral to them.

This book offers a way of thinking about building that can become a mental map upon which the co-ordinates of any knowledge or information relating to ecologically sustainable building could be included. The model is reflected in the structure of this book and is based on fundamental aspects of life on Earth. The model is intended as a way of understanding the knowledge available, not as an attempt to displace it. The elements of this model, the structure of this book, our pattern that connects, the very beeswax of building ecology are the issues of interdependency, thermodynamics and change.

This approach is based on new advances in the life sciences, particularly systems ecology. Chemistry, ecology and principles of thermodynamics are coupled with systems ecology, providing a broad context of general knowledge of building-related environmental science. We can use this expanding area of scientific knowledge as a basic field of inquiry as our aptitude in ecologically sustainable building develops during our lives. The key to understanding ecology is the knowledge that all elements of the system, whether living or non-living, are interdependent. Understanding interdependency shifts the emphasis for learning from the components of the system to the relationships between the components, from the parts to the whole.

These shifts in perception, based on an understanding of interdependency, also provide the basis for extending the concept of self-interest from our body and the building to include our greater 'self' of community, ecosystem and planet. Extending the notion of self-interest implies that we also know how to take care of our greater self – the ecosystems and life cycles of nature, and the communities of humans that, as building professionals, our actions affect.

Because of self-interest we have learnt which food is good for us and which is bad, what to do when we get too hot or too cold, and who to consult when we get sick. We know how to stay healthy. When we go to a pharmacy to buy a cure for a headache there are many remedies to choose from. If we choose the wrong one it might have no effect or only treat the symptom – the pain – rather than address the problem that is causing the pain. Being able to choose a remedy that will cure us depends on how well we understand the cause of our problem.

The building industry has over the years had a profoundly negative influence on the Earth's ecological health, largely because the economic system within which it now operates has ignored its interdependency with nature.[2] Many principles and strategies for building 'green', designing 'ecologically' and constructing 'sustainably' have been developed. Like choosing medicine at a pharmacy, understanding what building decisions cause ecological ill-health is an essential prerequisite for being able to either choose the best 'off the shelf' building remedy or for developing even better ones. By understanding the current scientific view of how nature works and how buildings link with and affect nature, we can gain a foundation of knowledge for understanding how to take action as building professionals to keep our greater self – the Earth, healthy too.

This knowledge, together with the technical ability and confidence to use ecologically sustainable building approaches, is the basis of ecological literacy. There are many building professionals all over the world who are applying this quality of ecological literacy to create ecologically sustainable building projects. In this book these people are called *building professionals* who are *ecologically literate* and *environmentally aware* (BEEs). They are the vanguard of innovation in the building industry because they are not only pioneering new building technologies and designs, they are also profoundly changing the economics of the industry so that it enhances rather than damages ecosystems. These BEEs produce the honey!

Presented in this book is the basic knowledge required to become a BEE and the laws and principles of ecologically sustainable building that guide their decisions.

What do BEEs know?

Our ecological literacy is a function of our understanding of required knowledge, coupled with our technical ability and willingness to use what we know. An ecologically literate building professional not only knows how to design, construct or manage buildings that contribute to ecologically sustainable development, they are also confident enough to *act* on the basis of their knowledge. Knowledge, technical ability and confidence are personal attributes that are gained over time. The information, case studies and activities presented in this book are intended to help facilitate gaining an initial level of ecological literacy, and introduce conceptual tools that can help us continue to develop our ecological literacy throughout our lives.

An important aspect of ecological literacy is understanding the environmental implications and effects of the different decisions made throughout the life cycle of a building development. Building-related environmental damage does not happen by itself. Environmental damage occurs due to the outcomes of decisions made by people responsible for different aspects of a project. Because environmentally literate people understand the likely environmental consequences of their decisions, they are empowered to make decisions that are environmentally beneficial.

Today's building industry leaders are individuals who not only participate in creating projects that have low environmental impact in terms of materials, processes and operational energy consumption. They conceive, nurture, promote and facilitate the kind of changes in building practice that are necessary to contribute to sustaining our life-supporting environments. BEEs have five essential types of knowledge that facilitate their ability to do this.

Knowledge of interdependency

BEEs know that buildings, built environments and the process of building depend on nature for all of their resources and that nature provides services like waste disposal and remediation that keep living systems healthy. They also know that what they build establishes flows of material and energy that affect present and future environments and people. This is the knowledge of interdependence.

Knowledge of conservation and efficiency

BEEs know that what goes around comes around – that matter is neither created nor destroyed but is continually circulating through all living and non-living systems. The first law of thermodynamics teaches us that energy can't be created or destroyed. This law focuses attention on how efficiently we use our materials and energy. As we cannot ultimately make any more of many of the resources we require for building, BEEs ensure that materials and energy are not being wasted, that as little material and energy is being used to solve as many problems as possible, and that we are doing more with less. This is the knowledge of conservation and efficiency or the first law of thermodynamics.

Knowledge of surviving designs

BEEs know that everything they build creates a system that continually needs the input of energy to keep it from breaking down. With this knowledge they can create systems that rely on renewable energy, and which use *energy-quality* in the most efficient and effective ways. They ensure that energy is used in a large number of small steps rather than in a small number of large steps. They create systems that use the outputs of consumption as resources for production, turning waste into food. This is the knowledge of surviving designs based on the second and fourth laws of thermodynamics.

Knowledge of natural systems

BEEs know that life is sustained by the constant cycling of materials from the Earth, through plants and animals, to the atmosphere and back through the Earth. They know that these grand cycles are what drive ecosystems and sustain a life-supporting biosphere. With this knowledge they ensure that their decisions address the imbalances in biogeochemical flows caused by human activity, and that their buildings help support the diversity of, and biodiversity within, ecosystems. This is the knowledge of natural systems.

Knowledge of change

BEEs know that the only certainty is that conditions always change. They understand that change is necessary for life to exist and that the ability to correctly perceive changes that are taking place is essential for adapting to new conditions. They know that the diversity of life and resilience in an ecosystem are the keys to its adaptability and that this diversity can be emulated when considering how a building functions over time. With this knowledge they know that a sustainable building is not one that must last forever, but one that can easily adapt to change. They apply their life-cycle thinking to create buildings that are resilient to environmental conditions and can cater to a diversity of human needs. In this way they ensure their buildings protect biological diversity, and minimise resource consumption and waste by avoiding obsolescence. This is the knowledge of change.

What do BEEs do with what they know?

Building professionals shape life. The built environments that are the products of our work become the systems that organise what we do and how we do it. The decisions that are made during the act of building, that is, choosing a configuration of spaces and elements and putting them together, affect environments that reach far beyond the physical site of construction and into the future. Because of rapidly increasing urban populations, our desire for built form is increasing and, as such, so is the ecological effect of the act of building. The building and the built environment represent complex human-created systems that are interconnected with other human and non-human systems, often of greater complexity.

As ecology is the study of relationships between organisms and the environment, building ecology is the study of the relationships between the act of building, the buildings and built forms that are produced, and the natural environment. Building ecology is the study of how our built home affects our natural home. It is about discovering the interconnections between buildings and nature and the effects of these interactions.

Building ecology is the study of the interdependencies and influences of building and natural environments on each other. It seeks to understand how natural systems both affect and are affected by buildings and construction by uncovering the relationships between them. The objective of building ecology is to discover ways of creating harmony between building and nature so that mutually beneficial and life supporting relationships can be designed and constructed.

In this regard building ecology can be thought of as a sub-set of the field of human ecology, a field of study that seeks to understand the interaction of humans with other species and their non-living environment of matter and energy.[3] A knowledge of building ecology is fundamental to understanding how built systems interact with natural systems. This is an ecological understanding. An ecological understanding is necessary to manipulate and recreate the relationships between buildings and nature so that the outcomes are positive and non-destructive.

The manipulation of relationships to create a desired outcome can be described as 'management', and can be described as building economics. *Eco* is derived from the ancient Greek *oikos,* means 'house'. Building professionals build and draw upon relationships between 'houses' and natural environments. *Nomos,* the ancient Greek for 'manage', also means 'steward'. Therefore, building professionals could be described as responsible for managing the effects of these interactions in ways which look after particular values and environments for someone else. In order to know how to manage a home we need to understand how it works. Ecology (*oikos* – house; *logia* – study) is the study of how our home in nature works, and is therefore prerequisite knowledge for good economics. In order to protect and enhance ecosystems, the holistic interaction of built environments and natural environments must be understood.

Once we have gained this knowledge it is important to know how to apply it in order to create buildings that are ecologically sustainable. As our knowledge of the relationships between building and nature improves we will discover nature's non-negotiable laws, and approaches to building that are essential in order to adhere to these laws and make decisions that lead to the best outcomes for people and the environment. These approaches are described in principles of ecologically sustainable building.

Scope of Building Ecology

Because built and natural environments are infused with each other, connected by mass and energy flows at many different scales, the relationships we discover and how we perceive them to be influencing an environment depend very much on how much of the system is being observed. It is therefore important to be clear about the scale of systems being considered in this book.

Introduction

This book does not provide detailed information on specific approaches to ecologically sustainable building, or the environmental impacts of different building materials. There is a body of existing knowledge on these issues and, throughout the book, readers are referred to appropriate publications and sources for further information on different scales of relationships and effects between the building industry, the broader urban and city scale, building procurement, landscape, building scale issues, building materials, and the scale of specific environmental impacts.

Each of these scales and the environmental effects of the activities considered within each scale are interdependent. Each scale affects the others. The questions that this book tackles are:

- How are these scales related?
- Is there a common basis for explaining why these scales are related?
- Is there knowledge available that can help students of building professions understand the origin and effects of issues common to all scales?

Answering these questions requires discussion across scales. Chapters therefore provide discussion relating to building materials, design and construction processes, neighbourhoods and cities, landscape, bioregions and global biogeochemical cycles. There is discussion of building life cycles and material life cycles, the metabolism of buildings and cities, as well as the carrying capacity of ecosystems.

These discussions aim to clarify the common properties of relationships that exist between all of these scales, the influence of thermodynamics on how systems develop to link scales, and the influence of time and change on building. With a better understanding of the workings of natural systems, we can create built environments that improve the integrity of Earth's ecosystems rather than contribute to their systematic destruction.

There is now widespread concern for the health and sustainability of the natural environment, and awareness of the ways buildings affect human and environmental health is now widespread. The concept of ecologically sustainable development has, in addition, spurred the development of government environmental policy and regulation, while the introduction of international standards for environmental management have set new benchmarks for assuring professional accountability for environmental protection. *Building Ecology* is therefore essential knowledge for those of us who want to be entrepreneurial and innovative with ecologically sustainable building, while meeting increasingly stringent *minimum* environmental performance standards.

A building is neither ecologically sustainable nor ecologically damaging of its own accord. Decisions made by people who create it decide its (and our) fate. Principles of ecologically sustainable building describe types of knowledge we require in order to productively and profitably relate to, and make decisions in the emerging environmental paradigm. However, knowing what to do will not necessarily lead to action being taken. What is essential is that we feel compelled to apply this knowledge in our professional practice. This requires not only technical knowledge, but the continued development of an environmental perspective applied to decision-making. This book therefore also offers an approach to developing *ecological literacy* so that people working in the building industry become BEEs.

The need for BEEs

The effect of building on our ecosystems has been predominantly negative. This is principally due to the structure of the relationships industry has with nature that take but which rarely replenish resources. BEEs are now working to create buildings that replenish the natural systems that they affect. BEEs try to make buildings that work more like trees do.

Trees and buildings have many things in common. They are both structurally strong, they both provide homes and workplaces for living things; both consume large quantities of natural resources and they both release emissions to our atmosphere. Trees and buildings can also be extremely inefficient.

A single walnut tree, for example, produces thousands of seeds, though only a few new trees may grow. In addition, the walnut tree discards thousands of brown leaves, covering its surroundings with its own detritus. In the process the tree consumes large quantities of water and nutrients. Such a large consumption of resources for such a meagre output! How inefficient. Buildings too consume vast quantities of natural resources at all stages of their life cycle, they output materials and emissions to their surroundings, and as an entity unto themselves, produce very little in return. The big difference is that the inefficiency of a tree is *sustainable*, while the inefficiency of a building is not. Why? Because, although both trees and buildings are often *inefficient*, only the human-made structure creates waste.

All of the consumption outputs of trees are, in one way or another, fed back to support production, replenishing the resources required to create more trees. Very little of the outputs from consumption involved with the life cycle of a building are kept 'in the loop'. Our systems of consumption do not support continued production. So, while we continue to design and construct buildings that don't take a leaf out of the tree's book – no waste, no pollution, resource-creating – then even creating buildings for maximum resource efficiency will only slow rates of environmental degradation. When buildings are created so that the consumption of resources produces outputs that support production of resources, the building industry will be making a positive contribution to ecological health.

So, in order to move from our current wasteful and inefficient present, to an effective resource-producing future, what do we need to learn? How best should we be taught? Most important, how might we be encouraged to apply an ecological understanding in our professional practice? Walnut trees have been growing and regrowing for millions of years in their natural environment. Will the built environment of our urban future allow us to continue to prosper? How might we begin planting the seeds for the 'next' industrial revolution?[4]

One key necessity is the creation of cultural and scientific relationships with nature that understand, value and protect the factories of life, ecosystems, and the development of social shapes and economic organisations that support these relationships. Buildings provide the physical infrastructure for our life, and they shape our lives in many ways. As Richard Register, founder of the global EcoCity movement, put it:

> 'We teach how to build, but what we build teaches us how to live'.[5]

Currently our built environments help us pollute, consume finite resources, destroy farmland and wilderness and overload water catchments. Even if we made our current way of living as efficient as possible, by reducing consumption and minimising waste and pollution, we would only prolong the inevitable loss of productive ecosystems and, concomitantly, human opportunity and peace. This is because the built systems that we live in are still consuming non-renewable resources, creating waste and creating pollution.

The World Resources Institute has documented the accumulation of the effects of patterns of development pursued by humanity, largely after the industrial revolution of the eighteenth and nineteenth centuries. In 1992 it reported that the resource base of this planet has reached a critical stage of degradation in three areas:

- erosion of the global soil base, reducing the world's capacity for food production as populations rise;
- loss of forests and wild lands leading to loss of biodiversity, threat to indigenous cultures, and degradation of slopes and watersheds;
- accumulation of pollutants and greenhouse gases in the atmosphere, leading to local hazards to soils, vegetation and human health, and the threat of global climate change.[6]

Buildings worldwide consume about 40% of the planet's material resources and 30% of its energy.[7] The construction of buildings is estimated to consume 3 billion tonnes of raw material per year, and generates between 10% and 40% of the solid waste stream in most countries.[8] The manufacture of many of the materials used in buildings require the consumption of large amounts of energy derived from fossil fuels, and the displacement of mega-tonnes of earth during mining. For example, for every tonne of cement, the world's most used building material, about two tonnes of raw material must be mined, nearly one tonne of CO_2 and up to 6 kg of NO_x greenhouse gas produced.[9,10] Building therefore contributes significantly to global ecological degradation and greenhouse gas emissions.

Industrial and economic systems established to meet our increasing demand for material goods do not account for the pollution and waste produced (sometimes in greater quantities than the products themselves) throughout a product life cycle, nor do they adequately consider the loss of ecological systems caused by resource extraction. Increased globalisation of trade has opened up new markets and financial opportunities, while exposing more of the Earth to resource depletion and pollution.

Due to the globalisation of trade, people in developed countries cause impacts on ecosystems that are not just hidden, but which occur in other countries. Many common materials and products are made up of elements which each come from different areas of the world and all of which have associated environmental impacts, as the following example from the World Resources Institute demonstrates.

'An urban professional in Tokyo reads a newspaper printed on pulped trees from North American forests. Her food and clothing come from plants and animals raised around the world – cotton and cashmere from Asia, fish from the Pacific and Indian oceans, beef from Australian and North American grasslands, fruits and vegetables from farmlands on four continents. The coffee she sips comes from tropical Central American plantations but is brewed with water from local wells.'[11]

The impact on ecosystems where these products come from is not the only consideration. When we begin to think about all of the transportation required to supply the products listed in this example, we begin to see that fuel extraction, processing, shipping, refrigeration, warehousing and roads all impact ecosystems in one way or another.

Within this environment the building industry works to provide the built-infrastructure for our lives. In many ways the design, construction and operation of our built environments have helped disconnect us from an awareness of the ecological consequences of our consumption, our pollution and our increased access to resources. Before we can begin to remove threats to our ecosystems, it is necessary to understand the contribution of current building practice to these issues.

It is clear that as more and more people become city dwellers, the demand for building must increase, yet an increase in construction activity without the integration of lessons learnt about the environmental impacts of common practice will make things worse. If we want to survive our urban future there is no option but to build in ways that not only reduce environmental damage, but which improve the health of ecosystems and protect natural resources.

This realisation has prompted the establishment of major initiatives by building companies, governments and communities worldwide that have aimed at least to decrease the environmental harm of building, and at best integrate new processes and technologies that lead to ecologically sustainable building. The years since 1992 have seen, for example, the establishment of environmental industry bodies like the US Green Building Council, the UK's Association for Environmentally Conscious Builders, the Australian Building Energy Council and similar organisations in countries like Canada, the Netherlands, Japan and South Africa. Many levels of government now use the International Standard for Environmental Management Systems, ISO 14000, or a local equivalent as a prerequisite for eligibility to tender on building projects and have sponsored many energy efficiency and building-related greenhouse gas abatement programmes.

Global research organisations like the Civil Engineering Research Foundation, International Council for Research and Innovation in Building Construction (CIB) and the International Energy Agency have sponsored many research programmes and conferences aimed at creating knowledge that can be applied to mitigating building-related environmental damage. The development of life-cycle assessment tools for the environmental assessment of buildings in design, and the availability of environmental performance rating schemes for completed buildings such as the UK Building Research Establishment's BREEAM[12] programme, are now common. At a community level there have been co-operative housing projects, use of organic materials, and community education programmes in sustainable building. This activity has been described as the movement for sustainable construction,[13] and there is little doubt that this style of building represents the desired direction of innovation for the building industry in general. However, despite the volume of sustainable construction activity and the availability of tools, techniques, information and education, ecologically sustainable building remains far from mainstream practice.

In the ten years since the 1992 World Resources Institute report, despite the initiation of *Agenda 21*, the *Kyoto Protocol* on greenhouse gas reduction and other global sustainable development and environmental protection measures including the movement for sustainable construction, the threat to life as we know it has increased. In its latest report the World Resource Institute warns that *ecosystems*, the Earth's primary producers and providers of life-supporting goods and services, are in decline:

> 'The current rate of decline in the long-term productive capacity of ecosystems could have devastating implications for human development and the welfare of all species.'[14]

Human society has evolved into a largely urban culture with security, shelter and sanitation being primary needs for our welfare. While the construction of this infrastructure leads to environmental damage, failing to provide these built amenities leads not only to environmental damage but also to social catastrophe.

Building is intimately entwined with ecosystems. It requires access to resources for energy and materials throughout its life cycle and it relies on ecosystems to assimilate waste and replenish resource stocks. While building professionals have always been aware of the need to have access to resources for energy and materials, the subtle natural mechanisms and processes for assimilating pollution and waste have remained largely unconsidered. Pollution and waste have therefore been exported, rather than eliminated, reduced or recycled. Until fairly recently, the effects of our resource consumption, waste generation and pollution have remained largely out of sight and out of mind.

In the mind of the BEE

Understanding the workings of nature has been perhaps the greatest mystery for the human mind. This great mystery has been the source of inspiration for artists and poets, of metaphor for philosophers, deities for religions, and fields of inquiry for scientists. Perhaps at the core of this mystery is one fundamental question: How is it that our body, the vehicle for our physical lives, is so obviously connected with the natural forces that surround us, yet our mind is seemingly unbridled? Why do we see ourselves as separate from nature when clearly we are not? Perhaps the answer lies in the way we have constructed our surroundings.

Our first buildings were created in response to the vulnerability of our physiology, to protect us from weather and from physical attack, to store food and prepare meals. Having achieved security, humanity began to construct buildings not only for the body but also for the mind. Churches for religion, houses for government, institutions of learning, and offices for commerce are all physical buildings that are more representative of human ideas like spirit, power, intellect and money, than they are with our physical interdependency with natural elements such as earth, air, fire and water.

The kinds of ideas that these buildings house have led to the creation of completely metaphorical constructions. The architectures of 'ivory towers', 'power structures', 'glass ceilings', 'corporate ladders', and 'trade barriers', for example, are not physical in nature at all, yet are assigned enough reality to motivate all kinds of human activity. Where once buildings were constructed to protect us only from the workings of nature, women's refuges, fall-out shelters and prisons are now constructed to protect us from the physical implications of our ideas. Many buildings are celebrations of human ideas, yet in building physical infrastructure for the activities of our minds, the need to look to the non-human systems of nature that support the functioning of our bodies and societies has very often been forgotten. No matter what motivation lies behind the creation of a building, no abstraction of the mind can alter some basic facts:

- It will be connected in some way to Earth;
- It will depend on nature for resources;
- It will cause environmental change;
- It will affect both human and non-human life.

A vast quantity of research shows that the accumulated effects of human action, particularly since the industrial revolution, have been undermining the functioning of the planet's providers of life-supporting goods and services, ecosystems. The physical effects of disruptions to these life-support systems, such as acid rain, ozone depletion, global warming, water pollution, species extinction and resource depletion, have concomitant impacts in human society. Environmental problems are being felt globally as well as locally and have reminded human minds all over the planet that our physical, social and economic well-being is intimately connected with the well-being of each other and of nature.

As this awakening gradually spread through indigenous people, scientists, activists, politicians, business leaders and through communities, human minds began to perceive the destructive implications of certain human activities and people were urged to 'save the planet' (from ourselves). Where once we built structures to protect our bodies from nature, armed with a new idea, we began to build in order to protect nature from us. Unfortunately this new idea is fundamentally flawed because it is not the planet that is at risk, but our existence and that of many other species and the habitats upon which life as we know it depends. Measures such as environmental protection laws, and agencies, as well as new technology and technological efficiencies, have not reversed the declining health of Earth's ecosystems.

Building 'efficiently' to limit the exposure of natural systems from destructive human activity, thereby protecting nature from us, does not deal with the root cause of our problems – that our relationship with nature has been a predominantly destructive one. It is obvious that a new idea for the human mind to build with is required. A change in the design of our building activity so that buildings support the functioning of ecosystems is now being pursued. Rather than building to protect ourselves from nature in order to protect our bodies, or building to protect nature from us in order to save the planet, we are now beginning to build *with* nature for the mutual benefit of 'our world'.[15]

The real difference between these types of building is in what our minds perceive our activity to be. We have over the millennia protected the body and created a world for the mind; the mind has perceived its world as separate from nature, and we have constructed our human environments in this manner. We are now beginning to foster a unity of mind and nature, an 'ecology of mind' as anthropologist Gregory Bateson[16] expressed it, that is, an understanding of how

16

mind and nature can be combined to create environments, both built and natural, that sustain life. In the building industry this unity is achieved by combining the minds of *building professionals* who are *ecologically literate* and *environmentally aware* (BEEs) with the understanding of nature gained through *building ecology*.

As we become BEEs we achieve this unity. When we do the world will never look the same again. This is because we will be able to understand the interdependencies and consequences of our decisions, no matter when they might occur. As Thich Nhat Hanh describes, we will be able to see the cloud floating in this piece of paper.

> 'There is a cloud floating in this piece of paper. Without a cloud, there will be no rain; without rain, the trees cannot grow; and without trees, we cannot make paper. If we look even more deeply, we can see the sunshine, the logger who cut the tree, the wheat that became his bread, and the logger's father and mother. Without all these things, this sheet of paper cannot exist. In fact we cannot point to one thing that is not here – time, space, the earth, the rain, the minerals in the soil, the sunshine, the cloud, the river, the heat, the mind. Everything co-exists with this sheet of paper … We cannot just be by ourselves alone; we have to inter-be with every other thing.'[17]

As a cloud floats in this piece of paper, so does building. The homes of the loggers, the roads, sawmills, the paper factory, the warehouses, retail shops and the office in which this paper gains its text are all here. Fundamental to making building ecologically sustainable is understanding the interdependencies, and recognising the *relationships between* our natural world and our built environment.

When BEEs look at a building they see the mines, the minerals and the forests from which materials are made, they see the roads upon which materials have travelled and the power-plants and refineries that supply the fuel for the journey. BEEs see the rivers that supply our water and which receive our run-off. BEEs see the atmospheric emissions caused by the production of the electricity running our building, and by the burning fuels used to transport people to and from it. Most importantly, BEEs see the demands on nature created by the choices we make and know how to make decisions that are life sustaining. In this book we will learn how look at a building and see clouds, rain, forests, ecosystems, people, the places our wastes go and how to make decisions that lead to ecologically sustainable building. We rely on all of this to provide what we need to build and to survive.

We will learn that when we build, we build relationships between people, and between built and natural environments. The outcomes of the relationships we create are manifest in the changes made to people's lives, the integrity of built environments, and the health of natural environments. We strive to always create positive change. While our buildings are intended as a means to this end, awareness of the relationships between building and natural systems has been lacking and as a result the effect of building in nature often overlooked. It is as though, in our haste to build, we have forgotten that our prosperity is coupled with nature's, that we are part of nature – not separate from it. In order to address this very problem we begin our education as BEEs by developing the most fundamental of all realisations: that everything is related with everything else. This is the knowledge of interdependency. In the next section we explore this knowledge.

References

1 Bateson, G. (1979) *Mind and nature: a necessary unity.* E.P. Dutton, New York. p.231.

2 Georgescu-Roegen, N. (1976) The entropy law and the economic problem. In: *Energy and economic myths: institutional and analytical economic essays.* Pergamon Press, New York. pp.53–60.

3 Diesendorf, M. & Hamilton, C. (1997) *Human ecology, human economy.* Allen & Unwin, Sydney, Australia. pp.xviii–xx.

4 McDonough, W. & Braungart, M. (1998) The next industrial revolution. *Atlantic Monthly* October, pp.60–66.

5 Register, R. & Peeks, B. (1997) Ecocity theory: conceiving the foundations. In: *Village Wisdom, Future Cities – proceedings of the third international ecocity and ecovillage conference.* 8–12 January 1996, Yoff, Senegal. Ecocity Builders, Oakland, USA.

6 World Resources Institute (1992) *World Resources 1992–1993* Elsevier Science, Amsterdam, Netherlands. In: Spence, R. & Mulligan, H. (1995) Sustainable development and the construction industry. *Habitat International* **19** (3) 279–292.

7 Roodman, D. & Lenssen, N. (1995) *A Building Revolution: how ecology and health concerns are transforming construction.* World Watch Paper No. 124, March. World Watch Institute, Washington DC.

8 Kibert, C.J. (2000) Construction ecology and metabolism. In: Boonstra, C., Rovers, R. & Pauwels, S. (eds) *International conference sustainable building 2000 Proceedings.* 22–25 October, Maastricht, Aeneas Technical Publishers, Netherlands.

9 Lawson, B. (1996) Building materials energy and the environment: towards ecologically sustainable development. Royal Australian Institute of Architects, Red Hill Australia.

10 Van Oss, H. (1999) Cement. *US Geological Survey Minerals Yearbook – 1999* Vol 1. Minerals and Metals. US Government Printing Office, Washington, USA. pp.16.1–16.13. http://minerals.usgs.gov/minerals/pubs/commodity/myb/. Accessed 4/02/02.

11 World Resources Institute (2000) *A Guide to World Resources 2000–2001: people and ecosystems: the fraying web of life.* Elsevier Science, Amsterdam, Netherlands. p.4.

12 Baldwin, R. (1996) *Environmental Assessment and Management of Buildings.* Building Research Establishment, Watford, UK.

13 International Council for Research and Innovation in Building and Construction (CIB) (1999) *Agenda 21 on sustainable construction.* CIB Report Publication 237, July. Rotterdam, Netherlands.

14 World Resources Institute (2000) *A Guide to World Resources 2000–2001: people and ecosystems: the fraying web of life.* Elsevier Science, Amsterdam, Netherlands. p.6.

15 Yeang, K. (1995) *Designing With Nature: the ecological basis for architectural design.* McGraw Hill, New York.

16 Bateson, G. (1973) *Steps to an Ecology of Mind.* Paladin, St Albans, UK.

17 Thich Nhat Hanh & Levitt, P. (eds) (1988) *The Heart of Understanding: commentaries on the Prajnaparamita Heart Sutra.* Paralax Press, Berkeley, USA.

PART I
INTERDEPENDENCY: HOW BUILDING AFFECTS NATURE

INTRODUCTION

'To see a world in a grain of sand
And a heaven in a wild flower
Hold infinity in the palm of your hand
And eternity in an hour'

William Blake[1]

The first kind of knowledge a BEE needs is the knowledge of inter-dependency. This is the knowledge of the relationships that either al-ready exist between building and natural systems, or that are formed during the building development process. By understanding these interdependencies, BEEs know why building affects nature and are therefore able to begin making decisions that form ecologically sus-tainable relationships. With a knowledge of interdependency BEEs are able consider the effects of their decisions on whole systems, rath-er than just elements of building like materials, finance or the site. As Capra writes, interdependency is the foundation for the development of an ecosystems approach:

> 'Based on the understanding of ecosystems…we can formulate a set of principles of organisation that may be identified as the basic principles of ecology, and use them to guide and build sustainable human communities. The first of those principles is interdependence.'[2]

In this section we consider three interdependent systems: the built-environment, biogeochemical cycles and ecosystems. Each is a dy-namic system and therefore changes in response to the influence of the others. We will discuss how our built environment influences both ecosystems and biogeochemical cycles. In turn, ecosystems change in response to biogeochemical and built-environment influ-ences. The relationship between these systems is depicted in Fig. I.1.

22

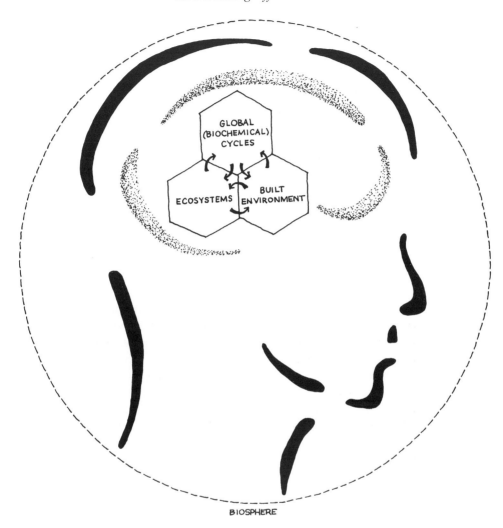

Fig. I.1 Trilogy of interdependency. Considering the interdependencies of ecosystems, built environments and global cycles is the basis of a whole systems approach to ecological decision-making. This way of thinking distinguishes BEEs from other building professionals.

General knowledge about each of these interdependent realms is provided but our focus is on is the *relationships between* each of them. Although we begin with a broad systems view of the relationships between building and nature, we must remember that the built-environment, ecosystems and biogeochemical cycles are elements of a yet greater system – the *biosphere,* and that the cumulative effects of the interdependency of these three realms (and others) dictate conditions for life on Earth.

23

Interdependency

This section consists of four chapters. Chapter 2 deals with life-cycle thinking. Ecosystems and urban systems are not static and our interdependency through time must be considered. Life-cycle thinking is a common approach to considering the accumulating impacts and effects of building activity through time. BEEs never make a decision without considering its life-cycle implications. We therefore begin our discussion of interdependency by exploring the concept of life-cycle thinking, as it is the fundamental perspective with which our discussion of interdependency takes place.

Chapter 3 describes urban metabolism. The concept of urban metabolism is introduced as a way of understanding the kinds of systems humans have built. We concern ourselves principally with the ways our urban systems rely on ecosystems and the challenges to ecological sustainability arising from the way our urban metabolisms function. Chapter 4 describes impacts on biogeochemical cycles and ecosystems being caused by current building practice. We then summarise the major interdependency issues and impacts in Chapter 5.

The metabolism of the building process and individual buildings will significantly influence whether a building will damage the environment, and to what extent. Because buildings can last a long time, the environmental implications of the resource inputs and emissions of building must be considered with a life-cycle perspective. We therefore begin by understanding life-cycle thinking – how BEEs consider interdependency through time.

2 LIFE-CYCLE THINKING: HOW BEES THINK ABOUT INTERDEPENDENCY THROUGH TIME

Introduction

Buildings are interdependent with nature over time. Relative to our own individual life-spans buildings as artefacts of human endeavour can last a very long time. Each building constructed can be regarded as infrastructure for the future as well as a historical record of our society, our economy and how we perceive our relationship with nature. From the day it is opened until well after those responsible for its creation are dead a building's design, materials, energy requirements, and its waste stream provides a built environment that people will construct their lives within and around.

Buildings like temporary exhibition facilities, movie sets and builders' site offices, on the other hand, last a relatively short time. Either way, what we require of a building today in terms of resource input and what we design its wastes to be, creates relationships with nature that can be more lasting than the structure itself. Building* therefore relies on the environment to always have the resources that are required to maintain it, and to be ecologically robust enough to assimilate the waste that building occupants produce. We can understand the relationships between a building and nature over time by thinking about its life cycle.

When determining the likely environmental effects of a building BEEs use the building life cycle to map the relationship between a building and the natural environment over time. The building life cycle refers to the stages through which a building development will

*Building is used in this sentence first as a verb (as in the act of building) and then as a noun (as in occupants of a building). This is a healthy confusion because it shows that BEEs concern themselves with ecological sustainability (a) in the act of designing and constructing a building and (b) in the life of the building they construct.

progress. The boxes in Fig. 2.1 represent these life-cycle stages. Each phase of the development life cycle involves different stakeholders, presents different responsibilities, and provides different opportunities for ensuring a building is ecologically sustainable. The roles and responsibilities of project stakeholders are discussed in more detail in Chapter 9.

Each phase of the building life cycle requires flows of materials and resources, and produces emissions to the environment. In modern building the natural environment is both a source of resources and a sink for emissions during all phases.

The flow of resources from nature to building and from the building to nature cause environmental impacts. The magnitude of the impacts is different at different life-cycle phases. Some of these environmental impacts accrue once, and others emerge over the life of the building. An example of a one-off impact might be the excavation of a basement during the construction phase. An example of a continuous and accumulating impact might be the building's contribution of greenhouse gas emissions as a result of the use of electricity generated by the burning of brown coal.

As a building has a life cycle, so does a building material. Construction materials are very often composites made up of many different types of raw materials. The mining, manufacturing, use, maintenance, reuse, recycling and disposal of every building material have important environmental ramifications. This story of a building's or a material's accumulation of environmental impact is called its life-cycle environmental performance.

Life-cycle assessment

Describing the environmental impact of a building can be thought of as telling a story about the combination of all of the environmental impacts of its constituent materials, of the construction process, of its operation and its refurbishment and eventual demolition. The complex web of interdependent natural systems in which impacts accumulate determine the effect a building has. Similarly, each material has a story that describes the origin of all of its raw materials, the resources required for processing, transportation, manufacturing, packaging, storage, delivery, use, maintenance, reuse, recycling or disposal.

Each story describes not only the resources required to make the building or material, but the environmental impacts at each life-cycle

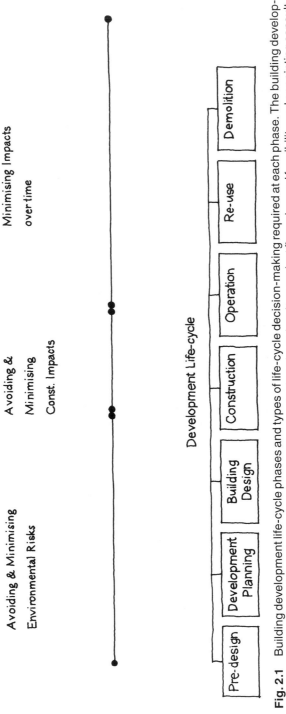

Fig. 2.1 Building development life-cycle phases and types of life-cycle decision-making required at each phase. The building development process proceeds through different phases. The pre-design phase involves project financing and feasibility and appointing consultants. Consultants work with clients during development planning to create master plans and obtain planning permissions from relevant authorities. Conceptual design is developed and then detailed during the building design phase. Tenders for construction are called for and let and the building built during the construction phase. The building is used during operation. During the reuse phase, buildings may undergo refurbishment to maintain their amenity or are altered to be used for a different purpose. When a building becomes obsolete it is often demolished.

stage. BEEs try to understand these stories as completely as possible. One approach that makes understanding these complex life-cycle stories easier is a process of evaluation called life-cycle assessment (LCA). An LCA considers:

- extraction, processing and transportation of raw materials;
- production, transport and distribution of resulting products;
- use, reuse and maintenance;
- recycling and final disposal.[3]

The environmental considerations at each life-cycle phase of a building and building material are represented in Fig. 2.2.

Life-cycle assessment (LCA) is a technique used to collect information about the environmental implications of the extraction, processing, manufacture, use and disposal of building materials and products. The European division of the Society for Environmental Toxicology and Chemistry (SETAC) describes the purpose of life-cycle assessment as a process designed to:

- evaluate the environmental burdens associated with a product, process or activity identifying and quantifying use of energy, materials and waste discharged into the environment;
- determine the impact of these resources and waste and their environmental discharges; and
- evaluate and put into practice, opportunities for improvement.[4]

Manufacturing companies including BMW and Philips use LCA to re-engineer many of their component and manufacturing operations in order to increase resource efficiency and reduce ecological impacts caused by their products. For example, both Philips and BMW now make products that can be disassembled when obsolete. This allows some components to be re-used, and others to be recycled, thus reducing waste.

Building product manufacturers such as Australia's BHP Steel, Boral and James Hardie have now applied LCA to some of their building products and have compiled databases which detail their life-cycle environmental loads.[5] Life-cycle assessment has also become an important marketing tool for the companies who can now tender for projects that require high environmental standards.

Life-cycle assessment is carried out in a four-stage process involving inventory analysis, impact analysis, effect weighting and implementation.

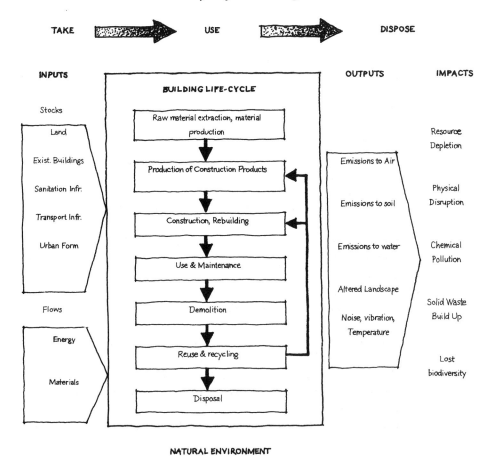

Fig. 2.2 Building life-cycle considerations. Life-cycle thinking requires us to consider the story of the accumulated environmental loading and associated impacts of a building material and building. Environmental loads and impacts are associated with the type and quantity of inputs to a material or building, the processes and use of material and building, the type and quantity of outputs from processes and use, and the nature of the receiving environments. Non-life-cycle thinking considers only how to take, use and then dispose of material, treating nature as merely the source of our materials and, after they are used, a sink for our wastes.

Inventory analysis is the first stage. In this stage all data relating to the physical properties of the product, and the environmental loads that it creates, are collected.

The second phase is *impact analysis*. At this stage the identified loads of a product are evaluated to determine the likely impacts over the product's life cycle.

The third stage is *effect weighting*. This involves determining the likely effects of these impacts and then weighting their importance, in order to indicate the relative environmental performance of the product and to decide what can be done to reduce its environmental impact. This third stage is perhaps the most contentious part of the process because the significance of impacts depends heavily on where and how the product is being used. Nevertheless, suppliers providing this information for building products are careful to include only data which is widely relevant, and to stipulate the scope of effects to have been considered in developing the weightings.

The fourth stage is *implementation*; during this stage courses of action, based on analysis of the LCA results, are taken to improve the product's performance.

Life-cycle assessments are now generally carried out using computer software, which contain regional information describing conditions upon which a product may have an impact. The programs compare these conditions with the physical properties of the materials in question and provide an output, which in many cases is a weighted index that indicates the relative environmental performance of the product. Life-cycle assessment procedure is now subject to standardisation under ISO 14040.

Example – life-cycle assessment in use

James Hardie, a large Australian building materials manufacturer, uses LCA to identify and mitigate environmental impacts associated with its products. In November 1999 it released an LCA report on its fibre-cement products.[6] Using Dutch LCA software and the company's own material input data, Hardies was able to map the life-cycle impacts of five of its sheet products. The LCA measured the products' relative performance on eleven impact categories and then normalised the results to show the contribution of different material constituents to the products' life-cycle environmental impact. The normalised results are shown in Table 2.1.

The LCA study found that cement contributed up to 35% of the environmental impacts, energy generation contributed 33%, with all other processes causing 32% of the products' impacts. Studies such as these help James Hardie to identify major environmental impacts and potential savings.

Table 2.1 LCA study parameters of James Hardie fibre-cement 6 mm Villaboard™ sheet.[6]

Environmental parameter	Unit	Amount/m² Villaboard™ Sheet
Embodied energy[a]	MJ	45.9
CO_2 air emissions[b]	kg	11.9
NO_x air emissions[b]	g	80.9
SO_x air emissions[b]	g	65.1
C_xH_y air emissions[b]	g	26.9
CH_4 air emissions[b]	g	2.4
Dust air emissions[b]	g	28.9
Solid waste emissions[c]	kg	6.8
Water resource depletion[d]	L	101.7
Wastewater discharge[e]	L	81.7

Notes: (a) embodied energy includes the energy content of all the process inputs and transportation energy; (b) air emissions include those of energy generation, inputs manufacture and transportation emissions; (c) solid waste emissions include those of the fibre-cement process, fibre-cement process energy generation and raw materials manufacture, but do not include the solid waste emissions resulting from the generation of required energy to produce raw materials or resulting from transportation; (d) water resource depletion relates to the manufacture of fibre-cement products, the fibre-cement process energy generation and raw material manufacture; (e) wastewater discharge is for fibre-cement process, cement and cellulose fibre production and sand mining.

Source: James Hardie International (1999) 'Towards Greener Sheets' an environmental profile of James Hardie fibre-cement products. *Research & Development Technical Bulletin*, November, Table 2, 6.

Because LCA is focused on materials, BEEs must use LCA tools in combination with other environmental performance techniques like *energy modelling* and *design guidelines* to create a picture of the impact of both the construction and operation of the entire building over its life cycle.

The environmental benefits of taking a life-cycle approach have been demonstrated on Stadium Australia, built for the 2000 Sydney Olympic Games. The project development team used LCA to optimise the project's environmental design beyond just enabling the choice of low environmental impact materials (see Fig. 2.3). Compared with a conventional stadium design, the final design for Stadium Australia saves 30% of annual primary energy consumption. The stadium design includes a gas co-generation plant that reduces greenhouse emissions by 37%, and rainwater collection and on-site water recycling that reduces water consumption by 77%. Most importantly, LCA enabled the project developers to meet stringent ecologically sustainable development (ESD) project performance requirements.[7]

Buildings and their materials are dependent on nature to provide resources and assimilate waste throughout their life cycles. It is important to remember this life-cycle interdependency and to ask whether the decisions you make will result in environmental improvement or degradation over time. The flow and storage of resources in built

Fig. 2.3 Stadium Australia. The design of the stadium for the 2000 Sydney Olympic Games was optimised for increased ecological sustainability using a life-cycle approach. Photography courtesy of the Sydney Olympic Park Authority.

environments, as well as associated environmental impacts, can be understood by studying their life-cycle metabolism.

References

1   Blake, W. (1803) Auguries of Innocence. In: Bronowski, J. (ed.) *William Blake: A selection of poems and letters.* The Penguin Poets series D42. Penguin Books, London: 1964.

2   Capra, F. (1997) *The Web of Life: A new synthesis of mind and matter.* Flamingo, Harper Collins, London. p.290.

3   Consoli, F., Allen, D., Boustead, I., Fava, F., Franklin, W., Jensen, A.A., *et al.* (eds) (1993) *Guidelines for Life-Cycle Assessment: a code of practice.* SETAC Europe, Brussels, Belgium.

4   Consoli, F., Allen, D., Boustead, I., Fava, F., Franklin, W., Jensen, A.A., *et al.* (eds) (1993) *Guidelines for Life-Cycle Assessment: a code of practice.* SETAC Europe, Brussels, Belgium.

5   James Hardie Australia (1999) Towards Greener Sheets: An environmental profile of James Hardie fibre-cement products. *Research & Development.* James Hardie International, Granville, Australia. November 1999.

6   James Hardie Australia (1999) Towards Greener Sheets: An environmental profile of James Hardie fibre-cement products. *Research & Development.* James Hardie International, Granville, Australia. November 1999.

7   Royal Melbourne Institute of Technology, Centre for Design (2001) *Greening the Building Life Cycle: Life cycle assessment tools in building and construction.* Environment Australia, Canberra, Australia. May 2001.

3 BUILDING METABOLISM: HOW BEES UNDERSTAND EFFECTS ON THE WHOLE SYSTEM

Introduction

Buildings, suburbs, towns and cities – in fact, any built area – can be thought of as being like a living body. It must be fed; it processes what it receives to keep it running and creates waste products that are released into the environment. Thinking of built environments in this way provides a mental picture of how buildings and bodies are interdependent with nature.

Built environments, like our bodies, have a metabolism, and like our body's metabolism, the metabolism of a built environment can be mapped and measured. One method for mapping the ecological interdependency of buildings and built environments is using the extended metabolism approach.[1] This approach helps map relationships between built environments and nature to indicate the health of the total system. The extended metabolism of a building, building development, urban area or city is described in terms of stocks, flows and impacts.[2] (See Fig. 3.1.)

In this chapter we will investigate the relationships with nature created by the stocks of resources we draw from to construct and maintain our built environment, as well as those resources that accumulate in the built environment. We then look at the natural resources that flow through our buildings to keep them operating. Many resources for both stocks and flows can cause negative environmental impacts. The impacts of our building metabolism are discussed in detail in Chapter 4.

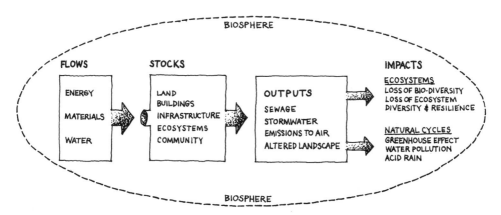

Fig. 3.1 Extended urban metabolism. The concept of urban metabolism provides a way of mapping and understanding the ecological interdependencies of buildings and built environments.

Stocks

The land, buildings, infrastructure and environmental health that exist prior to the commencement of a building project can be thought of as *stocks* or *inheritances* of resources. The community into which a building development is introduced can also be thought of in this way. BEEs make best use of available land and existing buildings and reuse materials from existing buildings to reduce resource consumption. Designs deal with limitations of existing infrastructure to protect environmental health. Where degraded environmental conditions are inherited, BEEs repair them. Where stocks of healthy ecosystems exist, no damage is done and protection offered.

Land

Only 18% of the Earth's surface is available for growing food, providing life-support services such as oxygen generation, terrestrial biodiversity protection, and carbon sequestration, as well as resource exploitation activities like mining and forestry. Built environments including roads take up a little over 2% of this available land.[3] The building industry predominantly uses land for building sites and landfill dumping of leftover building materials.

Land for building sites: from a BEE's perspective

In urban areas, land for building sites is scarce and thus existing buildings are demolished to make way for new construction. The availability of urban land suitable for new construction or redevelopment can also be affected by contamination caused by previous industrial or other uses. When urban areas have a consumptive and polluting metabolism, nearby ecosystems become degraded.

The easiest land to use for building development is cleared land that has never been built on. This land commonly exists on the fringes of urban areas. From an ecological perspective, building on previously undeveloped land in most cases involves replacing potentially ecologically productive land with ecologically unproductive land. In countries like Hong Kong and Singapore, where land for building is very scarce, marine environments are commonly 'reclaimed' for use as building sites. This urban encroachment into terrestrial and marine ecosystems reduces an area's biodiversity (see Chapter 5), another important stock.

The building construction process necessarily disturbs a site through activities like clearing vegetation, excavation, and general building works. Replacing topsoil with pavement also increases the rate of rainwater run-off and decreases the filtration of water through soils. Sediments and nutrients important to maintain soil structure and productivity are physically disturbed during construction processes and can be easily washed into aquatic environments where they can damage or destroy ecosystems.

The encroachment of urban areas on natural and rural landscapes continues to increase in some countries. In the UK, for example, an average of 6000 hectares of new, predominantly residential built environment replaces rural land each year, with the annual rate expected to increase to 6800 hectares by 2016.[4] In Australia 35000 hectares of land for new dwellings is expected to be required by 2005 to meet the growing demand for dwellings in non-metropolitan coastal areas.[5] These rates of urban encroachment are slow compared to that of many of the developing world's burgeoning cities.[6]

The land that builders of today are inheriting is also increasingly contaminated as the central concentration of industrial facilities in cities disperses to suburban and fringe areas, thus freeing inner city ex-industrial land for redevelopment.[7] Previously developed or contaminated sites for building are known as *brownfield* sites. Brownfield sites require remediation prior to redevelopment. There are estimated, for example, to be 450 000 brownfield sites in the US. Remediation of all of them would reportedly take approximately 75 years and cost $US750 bn.[8]

Land for dumping grounds: from a BEE's perspective

Land is also required for the millions of tonnes of solid waste accumulating, both within and beyond urban boundaries, each year. Australian urban areas, for example, dispose of about 2.5 kg of solid waste per person per day. With about two-thirds of Australians living in the five largest cities, that equates to approximately 32.5 thousand tonnes of solid waste per day for which disposal space is required.[9]

Unsustainable construction activity generates huge amounts of solid waste. Most of this material is dumped in landfills. In the UK approximately 800 hectares of land is taken up by landfill sites.[10] Landfills in urban areas are becoming scarce. In many affluent countries urban landfills are nearing capacity and new licences are not being issued due to community concern over noise, dust and odour, and because of higher environmental standards protecting, in particular, water courses and aquifers.[11] For land-scarce and densely populated countries, landfill availability is acute. As industrialisation and affluence of their economies increase, some countries are forced to take extreme measures to provide safe disposal for their waste. Singapore, for example, with no new land available for landfill, has constructed the Semakau Offshore Landfill (see Fig. 3.2), a new 350-hectare island, at a cost of $S610 m.[12]

With some countries having to resort to expensive projects like this one, coupled with increasing scarcity of landfill space, it is easy to see why the cost to the construction industry of continuing to use land for dumping waste is increasing. Projections from the US, for example, show that the average cost of depositing unsorted construction and demolition (C&D) waste at landfill sites can be as much as $US75/tonne.[13] Currently in Australia the cost is reaching around $A30/tonne. As landfill sites become less abundant and environmental controls become more stringent, costs will continue to rise and this will impact on the overall cost of construction and the market in general.

Fig. 3.2 Pulau Semakau – Singapore's landfill island. Land scarcity and increasing demand for waste disposal sites led to the construction of the new $S610 million landfill at Pulau Semakau. As this aerial view shows, the majority of the landfill is 'reclaimed' land. Photograph courtesy of the Ministry of the Environment, Singapore.

Example – increasing financial savings while reducing landfill dumping

Increasing tipping charges can provide opportunities to save money if the amount of solid C&D waste going to landfill is reduced. Construction companies include contingency sums for material wastage within their cost estimates. Any waste not created and therefore not dumped represents a direct saving to the project.

Civil engineering firm Baulderstone Hornibrook Engineering Pty Ltd (BHE), for example, saved approximately $A400000 per year by implementing a material recycling programme on the recent Western Link freeway project in Melbourne, Australia.[14] The project involved upgrading and widening 7.5 km of the existing freeway, building 4.2 km of dual elevated roadway and a 490-m balanced cantilever concrete bridge. How did they save so much?

- First the company's BEEs identified the major materials that were likely to be wasted. Construction materials include concrete, asphalt, and timber. Office materials such as paper and cardboard were also targeted for recycling.

- BEES at BHE then talked to waste contractors to determine the most effective methods for removing large quantities of materials from the site for recycling.
- Once the logistics had been worked out, a cost comparison between recycling targeted materials and disposal to landfill was conducted. Baulderstone compared the cost implications of the material type, travel distance and time, transport cost, tipping fees and recycling fees.

The firm found that, even when both landfill disposal and recycling incur a direct cost, recycling is usually cheaper because of an Environment Protection Authority levy on landfill dumping. The Western Link environment team's BEEs estimated, for example, that they saved about \$A1.50/m^3 by recycling concrete on the project. The results of BHE's economic evaluations show that recycling major waste streams was significantly less expensive than dumping waste in landfills (see Fig. 3.3).[15] Targeted waste materials were then either sent directly from the site to recycling facilities or were reused on the project.

Western Link was the first project to which BHE applied its waste minimisation programme.

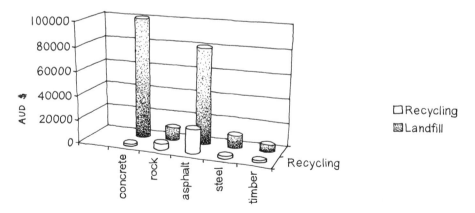

Fig. 3.3 Cost comparison of recycling and dumping on the Western Link Project, Melbourne, Australia, May 1998. The cost comparison conducted by contractor Baulderstone Hornibrook showed that recycling major construction materials in Melbourne was cheaper than landfill dumping. Reference: Master Builders Association of Victoria, EcoRecycle Victoria, Royal Melbourne Institute of Technology (1998) *The Resource Efficient Builder*, EcoRecycle Victoria Melbourne Australia Vol.1 No.3 May.

Existing buildings and infrastructure

Much of the advice available to environmentally aware building professionals about ecologically sustainable building focuses on new construction. New construction represents only a fraction of the total building stock. It is therefore imperative to deal with the issue of improving the environmental performance of existing buildings.

The BEE's perspective on existing buildings and used materials

Compared with many other human-constructed artefacts, buildings last a long time. As a result stocks of buildings existing today generally reflect the state of technology, economy and environmental awareness (and perhaps concern) of society in decades and centuries past. In this context, BEEs think of existing buildings as containers of useful space and stores of materials.

Containers of useful space

On this increasingly crowded planet, existing buildings provide stocks of an increasingly scarce resource – space. Clients often come to building professionals with a preconceived need for a new building, when what they are really after is a new use of space. With this in mind the first question a BEE asks is: Do you really need a new building?

Using space contained within existing buildings, rather than demolishing buildings to create new space, may greatly reduce the environmental burden of construction. For example, existing buildings with durable structures and flexible interiors can allow offices to become apartments, shop houses to become restaurants, or detached housing to become medical or commercial office space. This minimises resource consumption by avoiding demolition, saves energy by not having to reconstruct the structure, and in many cases helps preserve culturally significant buildings.

Wayne Trusty and Jamie Meil from the Sustainable Materials Institute in Marrickville, Canada, recently compared the environmental implications of building new versus renovating an existing structure.[16] Assuming complete interior retrofit, they found that where a building's structural system and envelope could be reused it would avoid:

- accumulating $1.67 GJ/m^2$ of embodied energy;
- the global warming potential of 0.45 tonnes of CO_2/m^2;
- the consumption of 1.25 tonnes/m^2 of resources; and
- the generation of approximately 120 kg/m^2 of solid waste.

Stores of materials

The building industry worldwide turns about 3 billion tonnes of raw material into buildings each year,[18] making for a lot of processed materials contained in buildings that could potentially be reused. Sweden's 3.1 million buildings, for example, contain approximately 130 tonnes of material per person,[19] while buildings in Osaka, Japan, contain 18.85 million tonnes of material or 2.2 tonnes per resident.[20]

BEEs try to draw as much as possible from the stocks of materials stored in existing buildings and as little as possible from nature. The recovery, reuse and recycling of materials is becoming more commonplace in the industry as natural resources become scarce and awareness of environmental impacts associated with resource consumption and the cost of landfill dumping increases.[21]

Example – making the most of existing buildings

Reusing existing buildings can be less costly, take less time, can retain cultural heritage and be more community oriented. The Singapore Housing Development Board's 'main' and 'interim' upgrading programmes for public housing estates provide good examples of the benefits of large-scale building reuse. Under the programmes, residents of housing estates erected 17 to 20 years ago are offered the opportunity to have their buildings upgraded in various ways. Works proceed if 75% of residents support the proposal and are prepared to share a percentage of the cost of works. External works include adding space through installing precast balconies, while interior spaces are reconfigured and services updated.[17] Upgrading of housing estates is made easier by their modular homogenous design, allowing new building elements to be precast and mass-produced off-site.

By upgrading rather than rebuilding, the pressure to find new construction sites in the land-scarce city-state is avoided, as is the environmental and financial cost of new construction, and the social disruption of relocating communities. When refurbishing existing buildings is not possible, the next option is to reuse secondhand building components and recycle materials that can't be used.

The BEE's perspective on sanitation infrastructure

Buildings in affluent cities are normally physically connected to non-built environments by sewer, stormwater and transport infrastructure. The configuration of this infrastructure bears greatly on the likely contribution of an individual building to environmental degradation.[22] Conversely, in the burgeoning cities of developing countries it is precisely the lack of sanitary infrastructure that leads to environmental and human health problems.[23] Sewerage systems have improved the health of people in cities remarkably. Unfortunately the centralised sewerage treatment technology used for the past 150 or so years has in many cities simply transported health problems to nearby watercourses and coastal areas.

Sewage infrastructure

The birth of 'modern' sewerage and stormwater systems occurred during the optimistic heyday of the industrial revolution at the beginning of the nineteenth century, when natural resources seemed inexhaustible and nature was regarded as something to be subdued and exploited. Although today we know that neither natural resources nor the ability for ecosystems to assimilate waste are limitless, the use of essentially nineteenth-century technology to deal with sewage persists when it comes to dealing with wastewater.

A typical large city is served by a sewerage system that consists of thousands of kilometres of pipeline connecting buildings to centralised treatment facilities. Vast quantities of water are required to move faeces, urine and, in many cases, trade waste through the labyrinth of pipes, creating a dangerous effluent. At a treatment plant, effluent is likely to undergo primary, secondary or tertiary levels of treatment, or will not be treated at all. The level of treatment determines its potential to damage the natural environment and its level of toxicity to people.

Primary treated effluent will be screened to remove large solids but will not be subject to disinfection prior to being released into the environment. Primary treated effluent is generally toxic and can pose a major health hazard to animals.

Secondary treated effluent will be screened and undergo some form of disinfection process (usually chlorination or ultraviolet irradiation) which kills bacteria but is unable to kill certain viruses. Secondary treated effluent in some cases can be used for irrigating playing fields and golf courses and as trade water. *Tertiary* treatment subjects effluent to further disinfection and is suitable for potable use, i.e. it is of drinking quality.

The final journey for a city's sewage is in many cases made along an outfall pipe. As most of the world's cities are coastal cities, the end of the pipe is likely to be in the sea (Fig. 3.4). Aside from the environ-

Fig. 3.4 Warning sign at outfall pipe, Victoria, Australia. This outfall discharges secondary treated sewage effluent into the surf near a popular holiday town on Victoria's west coast.

mental implications of effluent discharge, dumping millions of litres of useable water is a flagrant waste of a valuable resource.

The pipes and pumps of many cities' sewers are old, some exceeding 100 years of service, and are in need of constant maintenance. In countries like Australia, increasing population densities in inner urban areas are increasing the pressure on antiquated sewers. This means that the cost of running sewerage systems is becoming increasingly expensive.[24]

In areas where no sewerage system exists, buildings are likely to be required to connect to septic tanks. Simple septic systems merely contain effluent until it can be pumped out and taken to a treatment plant. Old septic tanks often leak effluent and contaminate groundwater. Innovative new septic systems can, on the other hand, provide highly treated effluent suitable for reuse. Some systems include ultraviolet treatment and provide treated water for irrigation of non-food crops. The main problem with either sewerage or septic infrastructure is that water is used as a medium of disposal (Table 3.1).[25] BEEs takes steps to minimise water consumption and to contain site run-off, thus reducing the load of a building on existing infrastructure.

Table 3.1 Metabolic flow of water through a typical Melbourne household.[25]

Location	Yearly volume of water in litres
Drinking water	270 000
To sewer	~175 500
(Port Philip Bay via Werribee)	~91 260
(Bass Strait via Eastern Treatment Plant)	~84 000
Stormwater from the roof discharged to drains and then via creeks and rivers into coastal waters (carrying with it oil, faeces and litter)	100 000
Total volume of wastewater discharged into Port Philip Bay, its estuaries and tributaries and the adjoining Bass Strait from Greater Melbourne's 1.4 million homes	315 801 242 000

Source: Newell, B., Hall, D. & Molloy, R. (1999) *Port Philip Bay Environmental Study Technical Reports.* CSIRO Environmental Projects Office, CSIRO Publishing: Melbourne.

Stormwater infrastructure

When BEEs look at downpipes and guttering they think not only about roof plumbing, but also the drains, streams, rivers and coastal areas that normally receive the rain that falls on buildings and building sites. In large cities this water flows through an extensive system of pipes and channels. Despite collecting any and all polluting substances that flow from city streets and building sites, stormwater receives little, if any, treatment. Most stormwater systems channel effluent into natural waterways.

Stormwater can be highly polluted. It has been stated, for example, that urban run-off contributes 37% of the world's oil pollution, with an annual volume of approximately three supertankers the size of the *Exxon Valdez*.[26] Pollution found in stormwater includes:

- oil leaking from vehicles;
- heavy metals leached from road surfaces;
- silt, fertilisers and pesticides from gardens and agriculture;
- bacteria and nutrients from animal excrement;
- bacteria and nutrients from sewerage overflows; and
- litter and organic debris from roads, footpaths, gutters and public spaces.[27]

It is important for BEEs that their building does not contribute pollution to these systems. They therefore try to use the water falling on their sites rather than letting it wash away, and apply techniques to improve filtration of stormwater leaving their sites. The threat of stormwater pollution is particularly high during the construction phase of a building project due to the level of site disturbance that occurs.

Stormwater need not be a pollution risk if it is considered as a resource rather than simply as something to be disposed of. In Singapore, for example, it is treated as a valuable resource. Singapore's catchment system includes natural and urban catchments areas. Stormwater in urban catchments is channelled to collection ponds and reservoirs before it is treated to high drinking water standards for supply to the nation.[28] This approach obviously reduces the run-off load on Singapore's aquatic ecosystems (Fig. 3.5).

The BEE's perspective on transport infrastructure and urban form

Transport infrastructure and urban form are other stocks with which buildings and their ecological implications are interrelated. The type of transport infrastructure present in an area, together with an area's urban form, largely determines the mode of transport occupants are likely to use.

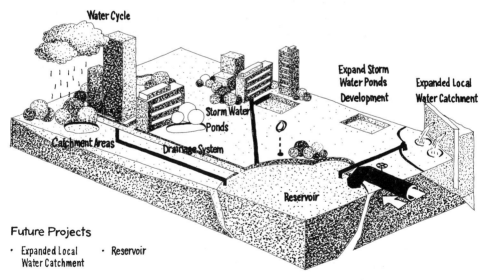

Fig. 3.5 Singapore's urban stormwater catchment system. Singapore plans to extend its ability to collect, treat and reuse stormwater, saving a precious resource and avoiding potential pollution problems. Figure redrawn from the original, courtesy of Public Utilities Board, Singapore and Silicon Illusions Pty Ltd.

45

Transport infrastructure

Transport infrastructure includes roads, rail and tramways, bicycle paths and walking trails. In areas located near or on waterways, water transport is also included. The mode of transport chosen by people to get to and from a building has a major influence on the quality of local and regional environments.[29] Table 3.2 presents data indicating that private cars contribute most to fossil fuel energy consumption, followed by public buses. The use of cars and buses also contributes greatly to air pollution. According to Newman and Kenworthy, rail transport is the most energy-efficient mass mode of land transport. Cycling or walking to and from buildings also needs to be catered for.

BEEs can also influence a reduction in the use of road transport during the construction phase of a building development. Transport of building materials in the UK is estimated to account for one-fifth of all freight, comprising approximately 14% of road and 5.7% of rail freight respectively.[30,31] Moving materials by sea is considered the most efficient mode of transport, with rail the next best option.[32] Unfortunately these modes of transport cannot be used to get materials to site. For this stage, road freight must be used.

With some major projects requiring huge labour forces, and vast quantities of materials to be transported for a number of years, BEEs discourage high-energy modes of transport and reduce the distance travelled by materials and people. The number of journeys to work

Table 3.2 Energy efficiency of transport modes for regional groupings of cities.[29]

Cities	Car (MJ/person)	Bus (MJ/person)	All rail[a] (MJ/person)
American	3.52	2.52	0.74
Australian	3.12	1.64	1.12
Canadian	3.45	1.61	0.51
European	2.62	1.32	0.49
Asian	3.03	0.84	0.16
Developing Asian	2.12	0.74	0.24

Note: (a) rail includes heavy rail and light rail and trams where relevant.

Source: Newman, P. & Kenworthy, J. (1999) *Sustainability and Cities: overcoming automobile dependence.* Island Press, New York. p.76.

taken by tradespeople in Australia's major urban centres, for example, was estimated in 1991 to total nearly 540 000 individual trips. Most destinations were outside of tradespeople's residential areas.[33]

Encouraging workers to car-pool or use public transport, as well as purchasing locally manufactured building materials, helps reduce transport requirements. This in turn reduces fossil fuel consumption, associated emissions and wear and tear on roadways.

Urban form

Urban form relates to the pattern of urban development and affects people's habits of movement and modes of transport. Newton[34] identifies six urban form archetypes. Most cities are a mixture of all of these categories:

- *Dispersed city* – formed by continued low-density development of housing and employment. Transport infrastructure is predominantly roads; main mode of transport is private cars (Fig. 3.6). This is typical of most Australian, North American and New Zealand cities.[35–37] This is the most energy-inefficient and polluting urban form. It has also been argued that this is the most inequitable form of settlement because it virtually forces residents to own cars.[38]
- *Compact city* – formed by increasing population density of inner suburbs and good public transport provision. Older areas of many

Fig. 3.6 New housing subdivision in Australia. How would you cope doing the shopping without a car?

Fig. 3.7 Toledo, Spain, a compact 'walking city'.

European cities take this form. The city of Toledo, Spain, for example, retains its compact medieval character and reminds us that
there was a time when streets were for people, not cars (Fig. 3.7).
The most compact urban forms are those that developed when the
major transport mode was walking. Areas of 'walking city' are also
being established in large cities. Within Melbourne, Australia, a
by-and-large dispersed city of 3 million people, areas where housing, employment and public transport nodes occur within a 1-km
radius (the maximum distance most people are willing to walk) are
being redeveloped as 'urban villages'.

- *Edge city* – formed by increasing population density, and employment at selected nodes. Orbital freeways are the major transport
  infrastructure, linking edge cities. The major mode of transport is
  again private cars.
- *Corridor city* – formed when urban growth is concentrated along
  corridors emanating from central business districts. These corridors are normally supported by public transport infrastructure.
- *Fringe city* – occurs when urban development is concentrated in
  suburban fringes of major cities. Urban developments on the suburban fringe are very often poorly serviced by public transport and
  car dependence is high.
- *Ultra city* – describes increased development in regional centres
  within 100 km of a major city. These areas become 'satellite' cities

linked to the nearby major city by fast rail links and freeways. The cities of Reading and Milton Keynes and their relationship with London in the UK are a good example of an *ultra city*.

While building developers may inherit transport infrastructure and urban form, buildings in turn have a major influence on transport patterns.[39] BEEs therefore use building design to influence the mode of transportation used by occupants. The main objective is to encourage less use of private cars and more use of public, bicycle and pedestrian transport. A key objective for BEEs is to provide a choice of transport options rather than providing only one. In the UK, for example, building designers are encouraged to provide secure bicycle parking, showers and change-rooms in office buildings to encourage employees to cycle rather than drive to work.[40]

Some housing developments in Europe are going a step further by discouraging residents from owning cars at all. They do this by locating near (or providing) good public transport, eliminating car access to the periphery of the development and scaling the development so that it is easy to walk or cycle around.[41,42] The integration of information technology into housing developments to encourage 'tele-working' (using the Internet to communicate with and work remotely from the central office) reduces the need for commuting and therefore the environmental consequences of transportation.[43]

Community: BEEs get involved

BEEs work with councils, community groups, churches, schools and conventional clients to establish building projects that enhance local as well as regional and global environments. Heightened community environmental awareness and increasing willingness to participate in environmentally positive activities is an important asset for ecologically sustainable building. BEEs know that mobilising and involving communities in development decision-making can be immensely beneficial for the long-term success of a building.[44]

Community awareness of and concern for environmental issues is high in many countries. Recent Australian Bureau of Statistics data indicates that 68% of Australians have a concern for the environment.[45] In the US the sustainability movement itself is considered a community driven, grass roots movement.[46] One of the indicators of community concern for its environment in Australia is the increasing participation of people in tree planting and land-care activities.[47] These activities are as much a part of urban as they are of non-urban environmental stewardship, and are important for fostering positive experiences of nature, particularly for youths in highly urban areas.[48]

In many developing countries people, rather than natural resources or financial capital, are the most abundant resource. Empowering communities to participate in housing provision is therefore a critical component of sustainable development strategies. The municipal government of Cape Town, South Africa, for example, is struggling to provide housing for rapidly increasing urban populations (Fig. 3.8). It has implemented a number of government-led strategies including building and renting council apartments (Fig. 3.9b), and encouraging small-scale commercial building contractors to build houses for those who can afford to pay (Fig. 3.9c). While these initiatives address provision of housing, they do not always meet people's social needs. In the Cape Flats area of Cape Town, women affiliated to the Homeless Peoples Federation, in conjunction with a non-government organisation, Peoples Dialogue, have created a community building programme that is creating better housing and community facilities (Fig. 3.9d).*

Fig. 3.8 The Cape Flats area, Cape Town, South Africa. Expanding informal settlements create social and environmental conditions that are a challenge to survive. Providing secure housing and sanitary infrastructure are essential for ecologically sustainable human settlements.

*I visited the Cape Flats area in November 1998, guided by officials from the Metropolitan Government of Cape Town. One of my guides now works at Planning South Australia and provided the details of the parties involved in the community-housing scheme.

(a)

(b)

Fig. 3.9 The results of different approaches to housing provision near Cape Town, South Africa. (a) Individual informal settlement house; (b) house built by private building contractors; (c) government provided housing; and (d) community-driven solution to low-cost housing. (*Continued.*)

Unfortunately the time it takes for community programmes to deliver housing here means that this approach alone is not sufficient

(c)

(d)

Fig. 3.9 (*Continued.*)

to meet the demand for homes required for the thousands entering South Africa's growing informal settlements (Fig. 3.9a).

The Eco-Centre project undertaken by the municipality of Port Phillip in Melbourne, Australia, is another example of community empowered building. The project involves the redevelopment of an existing house into a model sustainable building called the Eco-Centre, providing a community centre for learning ecologically sustainable urban living. When complete the Eco-Centre will provide both meeting and office space for community groups, and opportunities for the community to participate in urban sustainability projects. In addition an existing house is being refurbished into an eco-home. The refurbishment includes reconfiguring interior space to provide better access to natural light and enhance passive solar performance, and creating obvious links between the house, the garden, and Melbourne's first grey and black water on-site treatment and reuse system. This provides a small-scale opportunity for the community to learn about the impact of buildings on the environment and sustainable living.

Growing environmental awareness is also an increasingly common feature of professional and corporate communities. A professional responsibility for protecting the environment is now included in the charters of many professional organisations. The Institution of Engineers – Australia, the World Federation of Engineering Organisations, the International Federation of Consulting Engineers and the Royal Australian Institute of Architects have all recently published environmental principles by which they consider all professionals registered by them should operate.[49,50] Ecological sustainability has also become recognised as an ethical consideration. With increasing environmental regulation and demand for high performance buildings, becoming a BEE is also a financial imperative.

Ecosystems: a BEE's best friend

An oft-quoted axiom of the environmental movement is that our generation does not own the Earth, but is borrowing it from our children. Today's builders have inherited poor global, regional and local environmental conditions. The declining health of ecosystems is of particular concern then, not only for our generation of BEEs but of the BEEs to come. Preservation of functioning ecosystems is therefore crucial.

Functioning ecosystems are features of urban as well as natural environments. Stocks of both natural and urban ecosystems therefore need preservation. Endangered species and habitats are not just found outside city limits. They can also be found on or near development sites and therefore require protecting. For example, the green and golden bell frog, a species in decline near Sydney, was found to be living in wetlands in the path of the new M5 East motorway south of Sydney. The green and golden bell frog was lucky. Because of their ecological perspective the BEEs on this project were on the lookout for important species and habitats. The environmental team spent time collecting tadpoles from watercourses affected by the M5 project for relocation to a specially constructed habitat reserve located away from the path of the freeway. The new wetlands and a captive breeding programme were included as part of the project brief to ensure the frog's survival.[51]

BEEs sometimes incorporate biology or ecology into a project, using natural processes to replace technology. Protecting and incorporating existing ecology into building developments can help provide environmental services like storm and wastewater retention and treatment, and microclimate regulation at a far smaller financial cost than using human technologies that could do the same job. Sometimes it even pays to create habitat.

The Laurimar project in Doreen, Australia, for example, has constructed more than 4000 m^2 of wetlands (Fig. 3.10) and revegetated once derelict land with 57 indigenous species of plants, creating wetland habitat that had once been drained and cleared for agriculture. In this case stormwater retention and treatment is done by biology rather than technology while creating new habitat and increasing the biodiversity of the area. The wetlands provide a focal point for recreational activities and are a selling point for the new development. The BEEs on this project have recognised the interdependency of the wetlands, with the metabolism of homes, people's quality of life, and property value.

Flows

BEEs know that building is dependent upon flows of resources from our built and natural environments. Input flows from our environment are required at every phase of the building's life cycle. The environmental inputs required to construct and maintain buildings that have direct interdependency with ecosystems are energy, materials, air and water. These impacts can be positive or negative and are always interrelated. A building or built environment also relies on ecosystems to assimilate or store output flows of material considered to be wastes.

(a)

(b)

Fig. 3.10 Constructed wetlands at Laurimar, Doreen, Australia. (a) Aerial view of the site showing the Stage One constructed wetlands corridor between the two housing subdivision areas; (b) constructed wetlands provide aesthetic qualities as well as stormwater treatment. Photographs courtesy of Drapac Properties Pty Ltd.

Energy

> 'Everything is based on energy. Energy is the source and control of all things, all value, and all actions of human beings and nature.'[52]

In order to build a building, BEEs understand that energy must first have been consumed to produce the building materials and products, the construction equipment, and the transportation and storage infrastructure. This type of energy input is termed *embodied energy*. Energy is obviously also required to operate the building. This type of energy is termed *operational energy*. The environmental impact of a building is greatly influenced by the amount of *embodied energy* and *operational energy* it consumes over its life cycle.

Whether the discussion is about operational energy or embodied energy, BEEs know that the scale of impact will ultimately depend on consumption. The consumption of energy impacts on the environment, due largely to:

- the physical disruption caused by mining energy resources;
- pollution produced by energy conversion processes;
- greenhouse gas emissions; and
- dealing with wastes.

The best policy to pursue to avoid these impacts is to minimise the consumption of energy through efficiencies in building design and construction and through the use of low embodied energy building materials and components.

In Australia the energy consumed in construction is quite small in comparison with other industry sectors such as the electricity sector or road transport.[53] However, it can be seen from Table 3.3 that all other industry sectors, with perhaps the exception of agriculture, directly or indirectly input, or are related to, construction sector activity. One estimate put the proportion of delivered energy related to buildings in Australia at approximately 24% of total national delivered energy consumption.[54]

These figures demonstrate the importance of considering not only the environmental impacts caused by the operational energy consumption of buildings, but also the impacts associated with the embodied energy of materials and systems incorporated into buildings.

Most energy supplied in Victoria, Australia, for example, is generated by burning brown coal, the conversion of which to electricity is inefficient and produces significant quantities of CO_2 and other

Table 3.3   Energy consumption in Australia by industry.[53]

Industry sector	Energy used in Peta-Joules per annum[a]		
	1994–95	1997–98	2000–2001
Electricity generation	1147.6	1248.4	1323.8
Transport	1116.3	1182.9	1245.5
Manufacturing	1105.7	1210.7	1262.4
Residential	359.1	378.5	398.6
Mining	205.0	221.6	245.4
Commercial	176.1	192.4	209.8
Other[b]	67.8	71.1	74.5
Agriculture	63.5	67.3	71.3
Construction	44.2	48.0	50.7
Total	4285.3	4620.9	4890.8

Notes: (a) Peta-Joules (PJ) = 10^{15} joules; (b) includes lubricants, greases, bitumen, solvents and the energy consumed in gas production and distribution.

Source: Bush, S., Holmes, L. & Luan, H.T. (1995) *Australian Energy Consumption and Production: historical trends and projections to 2009–10.* Australian Bureau of Agricultural and Resource Economics. Research Report No. 95.1. Canberra, Australia.

greenhouse emissions. The energy consumption of existing buildings in Australia are responsible for generating about 80.9 Mt of CO_2 equivalent greenhouse gas emissions per year, which represents nearly 30% of total annual emissions.[55]

Embodied energy

The embodied energy of a material is the energy required to extract, process, manufacture and transport building materials and products and to deliver them to site. Embodied energy is a convenient indicator of a material's environmental impact due to the emissions and waste associated with the energy resources used in industry.

Embodied energy may be incurred more than once in a building's life cycle. Periodic maintenance and refurbishment add embodied energy to a building. An Australian study of the life-cycle embodied energy of a Melbourne office building found that, over a projected 40-year period, embodied energy accounted for 60% of the building's total energy requirements due to retrofitting and maintenance.[56] In Australia the amount of energy 'embodied' in existing building stock is equivalent to approximately 8.5 years of annual operational energy consumption.[57]

Table 3.4 Embodied energy for building materials.[58]

Material	Embodied energy (MJ/kg)
Air dried sawn hardwood	0.5
Stabilised earth	0.7
Concrete blocks	1.5
Insitu concrete	1.9
Kiln-dried sawn hardwood	2.0
Clay bricks	2.5
Plasterboard	4.4
Cement	5.6
Plywood	10.4
Medium density fibreboard (MDF)	11.3
Glass	12.7
Mild steel	34.0
PVC	80.0
Aluminium	170.0

Source: Lawson, B. (1995) Embodied energy of building materials. *Environment Design Guide*. Royal Australian Institute of Architects. PRO 2, August, 3.

The *absolute* amount of energy embodied in different building materials may vary depending on where and how different building materials are manufactured, and the environmental impacts of the production of this energy will vary depending on the fuel type and method of power generation. However, research has been able to provide us with a more accurate picture of the relative embodied energy of different materials. As Table 3.4 indicates, materials that require a high level of processing, or that require large amounts of raw material to be converted into a product, have a higher embodied energy than those materials that require less processing and can be used in buildings in, or near virgin form.

Material transport energy

The energy required to transport building material is normally included in embodied energy figures and is often regarded as only a small percentage of the overall embodied energy of a material. A number of recent studies have investigated the contribution of the transport of materials to the embodied energy of buildings. Miller[58] quantified the amount of transport energy required to deliver building materials to a 400 m² by seven-storey high apartment building in Brighton, England. He found that transportation of materials contributed 220.1 GJ to the embodied energy of materials, representing approximately 6% of the total. A similar study conducted on a $73\,m^2$ three-bedroom house in southeast London also showed transport energy as representing about 6% of embodied energy.[59] In both studies there were wide variations in the weight of materials, the efficiency with which they were loaded, and factors such as trucks carrying deliveries for more than one site, which meant that attributing transport energy-related impacts to specific materials was difficult.

Both studies therefore emphasised the significance of the distance a product travels and its mode of transport as most important in affecting the transport energy consumed and the associated atmospheric emissions. This means that the lower the processing energy a material requires, or the lighter the material, the greater the influence of transport energy on the material's overall embodied energy.

The percentage contribution of transport energy to the embodied energy of a material can increase with the distance and inefficiency of travel and as the weight and processing of a material decreases.[60] The transport energy for lightweight materials such as non-polymeric insulation and small steel components was 22% and >20% respectively, and contributed 55–70% of the embodied energy of low-processed material like sand and aggregate. Considering all of these issues, BEEs try to purchase locally manufactured materials to reduce transport-related embodied energy.[61]

Operational energy

While the embodied energy of a building is an important consideration, its influence on the building's energy-related environmental impact is not usually as large as the energy required to operate the building over its life cycle. Energy consumed by buildings in operation is significant. Buildings consume large proportions of a country's energy and, where generating fuel is coal or oil, contribute significantly to greenhouse gas emissions. Approximately 76 million residential buildings and 5 million commercial buildings in the US, for example, consume more than 30% of the country's total energy and 60% of its electricity.[62] BEEs therefore need to consider the fuel consumed to generate energy for their buildings as well as the way energy is consumed to operate them.

BEEs consider the operational energy requirement of the building important for a number of reasons. First, a building is typically constructed to be used for a very long time. As Fig. 3.11 illustrates, it takes between eight and nine years of operational energy consumption to equal the amount of energy consumed to manufacture the building's materials. If it is assumed that a building will be used for between 15 and 50 years, the amount of energy consumed to operate the building

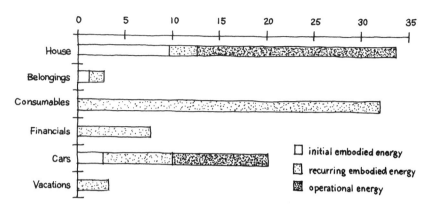

Fig. 3.11 Embodied energy of life. The amount of energy embodied in building materials, including energy added during maintenance and refurbishment (recurring embodied energy), in an Australian house is equal to about 12 years of operational energy. Given a 35-year life-span, the energy required for heating, cooling and appliances makes up the most significant proportion of life-cycle energy consumption. As houses become more energy efficient, embodied energy will become a more significant portion of total life-cycle energy.

could be up to five times the amount of energy required to build it.[63] BEEs therefore consider the long-term environmental impacts associated with demand for a particular type of energy.

Secondly, the amount of energy consumed during the lifetime of a building is greatly influenced by the habits of the building's occupants. For example, a building designed to be highly energy-efficient could end up being a huge energy consumer if people using the building don't know how to make use of its energy-efficient features. BEEs know that the lifestyle choices of building users can influence energy consumption beyond that required to actually run the building's services. Some buildings may help reduce the use of cars and thus fossil fuel energy consumption by providing facilities for cyclists like changing rooms, showers and bike racks.

Materials

It is obvious that the major input of materials into the building process is during construction. However, materials continue to be added and removed from buildings as they are maintained and refurbished during their lives. There are five factors that influence the ecological implications of materials used for buildings. They are:

- the magnitude and rate of material consumption;
- whether it is renewable or non-renewable;
- the toxicity of materials and how they are made;
- the sensitivity of the source of materials and how far they travel; and
- the wastes produced during material extraction.

Material consumption

If we choose to use materials that are being consumed faster than they can be regenerated then we are contributing to their depletion. Over-consumption of resources is thought by some to be the major problem that forces developed countries to address sustainability.[64,65] The amount of building materials used each year by the UK construction industry is equivalent to 6 tonnes of material per person.[66] Table 3.5 presents approximate quantities of major construction materials consumed globally. In 1995 the building industry was estimated to account for 40% of the annual consumption of raw stone, gravel, sand and steel, and 25% of virgin timber.[67]

The abundance of materials varies from region to region. While many building materials have been described as 'massively abundant'[68] they are not always abundant in the region in which they are most heavily consumed. Under these circumstances materials must be extracted from many different ecosystems around the world and transported increasing distances before arriving on site.

Table 3.5 Proportion of global material consumed used in construction.[67]

Material	% of global consumption used in construction
Raw stone, gravel and sand	40
Virgin wood	25
Energy resources	40
Water	16

Source: Roodman, D. & Lenssen, N. (1995) *A Building Revolution: how ecology and health concerns are transforming construction.* World Watch Paper No. 124, March. World Watch Institute, Washington DC, USA. Table 1, p.23.

How much is enough?

The term over-consumption indicates that more resources are being consumed than should be. We can't build without consuming resources. The question is then: How much should be consumed?

The amount of ecological resources required by a person to run their lifestyle has been called their 'ecological footprint', and a method for measuring the size of their consumption of ecological resources in land area, known as ecological footprint analysis has been created.[69] An ecological footprint analysis can be conducted to determine the amount of ecologically productive land required to satisfy the consumption patterns caused by decisions, including building development decisions.[70] It is also a way of measuring our ecological interdependency.

Ecological footprint analysis has shown that if the world's current population of almost six billion people were to live the same kind of high-consumption lifestyle as an average person in North America we would need three times the amount of ecologically productive land currently available on Earth. In other words, we would need three Earths to support our current population.[71] It is therefore impossible to sustain such lifestyles in the long run, particularly if fossil fuels remain the predominant source of energy.

The good news (if you can call it that) is that the current size of our ecological footprints is, in part, a function of the inefficient ways we use resources. For example, up to 70% of the fuel energy in brown coal can be lost during conversion to and transmission of electricity. When the electricity generated is used to power an incandescent light globe, only about 10% is converted into light; the rest is emitted as heat, making light globes far better heaters than sources of light! The Rocky Mountains Institute suggests that eliminating this sort of waste can contribute to four-fold increases in the efficiency with which we use resources.[72] However, achieving resource efficiency on this scale will not be enough to protect the viability of ecosystems because many of the ways we use resources cause pollution, waste, and ecological degradation. While resource efficiency is an essential ingredient in an ecologically sustainable approach to building, people in developed countries also need to change their high-consumption lifestyles.

Resource type: renewable, non-renewable or reusable?

An important factor affecting the sustainability of resource consumption is whether renewable or non-renewable resources are being consumed and how they are being consumed. Most of the materials we use in building are made from raw materials that are non-renewable. We also use materials like timber, which can be considered as renewable, in unsustainable ways. How then is it possible to build sustainably? On the other hand it is possible to use some building materials, like steel, derived from non-renewable raw materials, in ways that help conserve resources. Confused? BEEs aren't. Let's consider the matter more closely.

Renewable resources are those that can be used and reused without depleting their primary source. Solar energy, for example, is renewable because we cannot deplete the sun of its atomic energy. Wind energy is also renewable because we cannot deplete the source of wind (the sun), and tidal energy is renewable because we cannot deplete the gravitational pull of the moon.

Using renewable resources does not automatically lead to ecologically sustainable results. If a renewable resource like timber is being consumed at rates beyond which it can be regrown then it is not a sustainable resource. Trees may be able to regrow, but the forests they come from take hundreds or thousands of years to develop their biodiversity. Trees are a renewable resource but old-growth forests are not. The availability of a renewable resource can also be affected by pollution. Water, for example, is technically a renewable resource, however, due to pollution by industry, urban areas and agriculture significant quantities of available water are unsafe for use.[73]

The World Watch Institute offers an opinion of what constitutes a sustainable level of consumption of resources. In its guide to world resource consumption and production, *Vital Signs*,[74] the point at which consumption of a resource becomes unsustainable is when it breaches the 'sustainable yield' capacity of that resource.

Sustainable yield is usually understood as renewable resource harvest rates that equal regeneration rates.[75] But this understanding ignores emissions and other forms of pollution involved in extracting and using these resources, so our concept of sustainable yield must therefore be expanded to include all issues that affect the ability of the ecosystem to continue to function while providing resources and assimilating waste. This ability is termed the *carrying capacity* of an ecosystem.[76] Consumption levels which produce effects that are not within the carrying capacity of an ecosystem are therefore excessive.

The effect of resource consumption on both supplying and receiving environments can be reduced if resources are reused and recycled. The sustainable use of a renewable resource is therefore dependent on:

- reducing consumption to a rate equal to that which the resource can regenerate;
- avoiding consumption that causes ecological damage or pollution;
- reuse and recycling.

Non-renewable resources are those that are derived from depleting reserves of a primary resource. Minerals and fossil fuels are the two categories of non-renewable resources most heavily consumed. They are finite and their extraction is ultimately unsustainable. All fossil fuels can be considered as non-renewable because they contain energy in deposits that have taken millions of years to form. However, they contain large quantities of concentrated stored energy, are readily obtainable, and easily transportable and are therefore extremely useful.

We are using fossil fuels such as coal, oil and natural gas at a rate far exceeding that at which they are formed and are therefore depleting our stored energy reserves. Not all fossil fuels, however, are scarce. Discoveries of new reserves, coupled with increasing energy efficiency in energy consumption is likely to shield developed countries from the effects of resource depletion for some time yet. While natural gas, for example, is estimated to have 145 years of supply in current reserves, its availability is predicted to increase despite growth in global in demand.[77] Some metals and aggregates on the other hand are becoming scarce. Reserves of zinc, mercury and lead are expected to last merely 20 years.[78] The potential scarcity of zinc is of particular concern because it is currently used extensively in corrosion protection for steel.

Materials like metals and fossil fuel have become important resources for modern life, and our access to these resources has greatly improved human living conditions albeit at a high environmental cost. Many of the resources we have come to rely on are now scarce, requiring that we do more with less. Reusing and recycling materials avoids the environmental damage associated with resource extraction and also maintains reserves of natural resources. The level of consumption of non-renewable resources that can be sustained is therefore dependent on:

- reducing consumption;
- conserving scarce resources;
- the ability of a resource to be reused and recycled.

Reusable resources, used or secondhand materials and materials with recycled content, are also inputs for building and should be considered wherever possible to reduce the ecological burden the industry places on ecosystems through the use of new materials. Reusing and recycling materials helps increase the sustainability of using either renewable or non-renewable resources. Reuse of materials varies from country to country.

In developing countries reuse of materials can be near 100%. For example, faced with extreme scarcity of even the most basic building resources, people in the Cape Flats informal settlements near Cape Town, South Africa, have developed sophisticated salvaging and recycling operations. Figure 3.12(a) and (b) shows a recycling yard containing materials often salvaged from dumps rather than demolition sites.

It is developed countries that can afford the luxury of waste. Despite this, developed countries report quite high rates of reuse and recycling of materials. In Australia it has been reported that 80% of the materials contained in old timber houses are recovered and used for renovating and remodelling existing homes, while 58% of materials from commercial demolition are reused in other buildings.[79] Developed nations are beginning to create infrastructure and regulations to encourage the reuse of building materials. In Australia, for example, an online trading post for salvaged timbers called 'The One Stop Timber Shop' (Fig. 3.13) has been established to make it easier for specifiers, and builders to find out what materials are available and at what price.[80] A different approach in the Netherlands saw legislation passed in April 1997 prohibiting the dumping of reusable building waste. This has contributed to ensuring about 80% of materials in the Dutch construction and demolition waste stream are reused in construction.[81]

Material transportation

Building materials are travelling further. A recent case study of the origin of materials used to construct a Cypriot house built in 1994 found that while brick clay, bricks, sand, gravel, cement and facing stone were sourced from within a 50-km radius on the island, a large proportion of materials were sourced internationally. Used in the building were timber from the US, Belgian steel, Spanish and Italian tiles and various other materials from countries such as Japan and Russia. A building of similar design built at the beginning of the twentieth century sourced nearly all of its materials from within a 15-km radius.[82]

(a)

(b)

Fig. 3.12 People in developing countries can't afford to waste resources. (a) Recycling yard in the Cape Flats area, Cape Town, S.A. (b) Prefabrication in the Cape Flats area, Cape Town S.A. These wall frames are made from salvaged timber and metal sheet and, for those who can afford them, become the walls of 'informal' dwellings in sprawling squatter settlements.

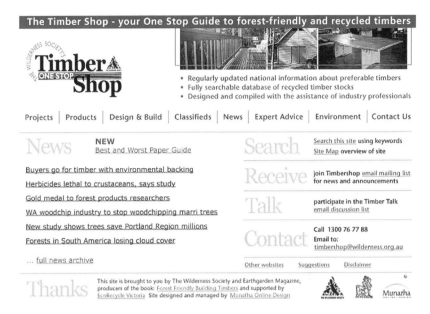

The Timber Shop - your One Stop Guide to forest-friendly and recycled timbers

• Regularly updated national information about preferable timbers
• Fully searchable database of recycled timber stocks
• Designed and compiled with the assistance of industry professionals

| Projects | Products | Design & Build | Classifieds | News | Expert Advice | Environment | Contact Us |

News

NEW
Best and Worst Paper Guide

Buyers go for timber with environmental backing

Herbicides lethal to crustaceans, says study

Gold medal to forest products researchers

WA woodchip industry to stop woodchipping marri trees

New study shows trees save Portland Region millions

Forests in South America losing cloud cover

... full news archive

Search
Search this site using keywords
Site Map overview of site

Receive
join Timbershop email mailing list
for news and announcements

Talk
participate in the Timber Talk
email discussion list

Contact
Call 1300 76 77 88
Email to:
timbershop@wilderness.org.au

Other websites   Suggestions   Disclaimer

Thanks
This site is brought to you by The Wilderness Society and Earthgarden Magazine, producers of the book: Forest Friendly Building Timbers and supported by EcoRecycle Victoria Site designed and managed by Munatha Online Design

Fig. 3.13 The One-Stop Timber Shop. Provides an on-line trading post for recycled and forest-friendly timber. Using the Internet in this way increases the ease with which customers and suppliers can communicate and provides a template for developing markets in used and ecologically sustainable materials. Reproduced with permission of the Wilderness Society of Australia.

Material toxicity

All materials are toxic in relevant concentrations and form. The level of toxicity differs throughout a material's life cycle. Lead, for example, has proven toxic when used in plumbing and paints. It is safe to use as flashing material but poses risks of poisoning again when it becomes worn or has to be removed. Many materials in common use are toxic to humans and the environment. A recent survey of chemical risks in the built environment identified a large number of dangerous substances used in the building industry. For example:

- cadmium and chromium used in electroplating are cancer causing;
- chromate and arsenic used in timber preservatives cause contact dermatitis on exposed skin and can cause lung cancer if inhaled;
- formaldehyde gas emitted from urea-formaldehyde foams, timber adhesives and reconstituted timber products like particle board and medium density fibre board (MDF) is an irritant and is a suspected carcinogen.[83]

Building material life-cycle assessment (LCA) programs are useful for determining the combined environmental effects of all manufacturing inputs and transportation of a material.

Waste associated with material extraction

Each year around the world billions of tonnes of material are mined and processed to make building materials. In many cases, the amount of material removed and refined vastly exceeds the quantity of finished building material produced. The volume of material moved and wasted during the production of a building material contributes to its environmental burden. This can be described as a material's *ecological rucksack*.[84] As shown in Table 3.6, the ecological rucksack of common building materials like cement can be up to ten times greater than the amount of raw material manufactured. In other words, it can take the equivalent of ten tonnes of limestone to make one tonne of cement. Similarly, it can take five tonnes of bauxite to make one tonne of aluminium, 14 tonnes of iron ore to make one tonne of iron, and 420 tonnes of copper ore to make one tonne of copper.[84]

BEEs use building materials with small ecological rucksacks and ensure that all materials are used efficiently. This helps reduce the overall ecological burden of the building project by minimising waste associated with material production.

Added to the avalanche of matter associated with the production of materials are the large volumes of building materials that are wasted

Table 3.6 Ecological rucksack of material.[84]

Material	Rucksack size in relation to one unit mass of the material
Bauxite	5
Cement	>10
Gypsum	5
Iron	14
Sand and gravel	>1
Zinc	27

Schmidt-Bleek, F. (1994) *Carnoules declaration of the Factor Ten Club*. Wuppertal Institute. In: Weizsäcker, E. von, Lovins, A.B. & Lovins, L.H. (1997) *Factor four: doubling wealth - halving resource use: the new report of the Club of Rome*. Allen & Unwin, Sydney, Australia. Figure 34, p.243.

during building construction or demolition and end up dumped in landfills.

Solid waste

The annual contribution of construction and demolition waste to the solid waste stream varies slightly depending on the level of construction activity taking place. Levels remain on average a very high proportion of the total solid waste stream in most countries. In Australia it has been estimated to constitute up to 40% of the urban solid waste stream disposed at landfill sites.[85] In Singapore it comprises between 10% and 20%,[86,*] while figures for the US are around 33%.[87] The building industry can therefore play a major role in trying to reduce solid waste and reduce the need to take up space for landfills.

This level of solid waste generation has significant financial as well as environmental costs. Reports from the UK suggest that the cost of disposing solid waste can eat up to 25% of a building company's profit margin. Taking an industry-wide view, achieving a 10–20% reduction in solid waste in the UK could potentially divert 6 million tonnes from landfill and save £60 m in disposal costs.[88] Australian building companies participating in a recent construction industry-wide effort to reduce solid waste in Australia, known as the Waste-Wise Construction Program, showed it was possible to reduce solid waste from construction projects by up to 50% compared to business-as-usual construction practice.[89] Table 3.7 gives comparative data for some European countries.[90]

What is wasted the most?

In March 1993 Hong Kong Polytechnic tabled their final report on reduction of construction waste in Hong Kong.[91] This report identified causes and quantities of waste produced on commercial projects at various stages throughout the building cycle. This information has been represented to show percentage contribution to total waste by construction phase and is shown in Fig. 3.14.

Recent research for the US Environment Protection Authority[92] characterised construction site waste on a number of different types of construction projects. The results show that in the US timber accounts for the majority of waste produced in residential construction accounting for about 42% by weight of total waste, with plasterboard

---

*Percentages of total solid waste stream in Singapore vary depending on whether metal and wood waste are added to construction debris figures or not.

Table 3.7   Annual construction and demolition waste in various countries.[90]

Country	C&D waste (million tonnes)	C&D waste (kg/person/year)
UK	30	509
Germany	59	720
Netherlands	11.2	718
Belgium	6.8	666
Austria	4.7	580
Denmark	2.7	509
Sweden	1.7	193
Finland	1.3	255
Ireland	0.6	162

Note: Figures do not include excavated soil, rock or road works.

Source: 'Core C&D Waste' Estimates from Member States, OECD and Study Team in Symonds Group Ltd in association with ARGUS, Cowl and PRC Bouwcentrum (1999) *Construction and demolition waste management practices and their economic impacts, Report to DGXI, European Commission, Final Report.* February, Figure 7.1, 44.

sheet the next highest contributor to on-site waste consistently accounting for about 27% of the waste stream.

Tenancy fitouts in existing offices generally produced significantly more plasterboard waste. Metal and timber studs were also a significant proportion of waste produced on the projects surveyed. Packaging of building materials has also been identified as a significant contributor to total waste produced during the construction of houses in the US accounting for a significant percentage of all wastes produced.[93]

Causes of waste

Studies have indicated that similar causes of building waste occur worldwide. Formoso *et al.*, for example, identify four possible times during a development when waste could occur, namely before the materials arrive on site; during transport, delivery, and storage; during production; and other events on site such as theft or vandalism.[94]

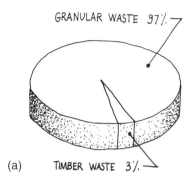

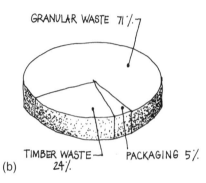

Fig. 3.14 Contribution to waste stream by building stage and waste type in Hong Kong. This data relates to quantities of materials wasted on concrete framed commercial buildings in Hong Kong. Granular waste refers to discarded concrete, sand, mortar and earth. Knowing what is likely to be wasted, when and in what quantities, is an essential basis for successful construction waste minimisation planning. (a) Waste in substructure: The high proportion of granular waste can be attributed to excavation and concrete substructure construction. Timber waste flows from shoring and formwork. (b) Waste in structure: Concrete is the predominant material used at this stage and accounts for the highest proportion of the waste stream. Timber waste flows mainly from formwork, while packaging of plant and equipment begins to become a recognisable waste stream. Reference: Hong Kong Polytechnic (1993) *Reduction of construction waste: final report*. The Hong Kong Construction Association Ltd, Hong Kong. Reproduced with permission of the BDP *Environment Design Guide*, Australia.

(See Table 3.8.[95]) Other research indicates that clients can be a source of waste through careless inspection procedures, variation orders or during part possession.[96]

A comparison of these studies shows that despite variations in construction style from nation to nation, potential material waste is caused by similar inefficiencies in design, procurement, materi-

Table 3.8   Causes of construction waste at different project phases.[96]

Design	Purchasing	Handling	Construction	Other
Plan errors	Shipping errors	Poor storage	Human error	Theft
Detail errors	Ordering errors	Deterioration	Equipment error	Vandalism
Design changes		Poor handling off-site & on-site	Poor housekeeping	Clients' actions
			ACCIDENTS	
			Weather	

Adapted from Graham, P. & Smithers, G. (1996) *Construction waste minimisation for Australian residential development Asia Pacific Journal of Building & Construction Management*. **2**(1), 14–18.

als handling, operation or residual on-site waste such as packaging (Fig. 3.15).

Water

Most fresh water is consumed by agriculture, accounting for up to 75% of total consumption globally. In Australia, agriculture consumes over 60% of fresh water resources. However, domestic use (use in and around buildings) is the second largest consumer. This represents 0.1% of global water consumption. Industry accounts for approximately 25% of global water consumption. There are regional differences though. In Singapore 53% of fresh water is consumed for domestic and municipal use[97] while 12% of the US's fresh water is consumed in houses and commercial buildings.[98] In order to take appropriate action to minimise water consumption and pollution, BEEs need to know where in buildings most water is used and how.

Toilets

The ubiquitous pedestal of human contemplation, the toilet is one of the developed world's major water consumers. When using a standard flush toilet in the US we are likely to consume more water than most people in developing countries use in an entire day.[99] The average Australian WC uses between 6 and 9 litres of water per flush, although buildings built after the introduction of mandatory code requirements of dual-flush cisterns use between 3 and 6 litres per flush.

An awareness of the inherent inefficiency of sewer systems and a desire to reduce water consumption has driven recent technical advancements in toilet technology. There is now a range of toilet systems

(a)

(b)

Fig. 3.15 Construction and demolition waste and job-site recycling. (a) Unsorted construction debris on a commercial building project in Singapore becomes a burden on scarce landfill sites. Photograph courtesy of undergraduate construction management student, Department of Building and Construction Economics, RMIT University, Australia. (b) On-site separation of waste streams for recycling can save money as well as divert materials from landfill. Here timber from a house deconstruction project in Oregon, USA, is sorted into a bin for recycling. Photograph courtesy of the Centre for Building and the Environment, R.E. Rinker School of Building Construction, University of Florida, USA.

that help increase the efficiency of water use, or that use *no* water at all and still provide adequate sanitation. Examples include low-volume cisterns that operate on a 6-litre/3-litre dual flush, split bowl pans that separate urine from faeces so that either or both can be used for fertiliser, and wet and dry composting toilet systems. Building-scale package treatment systems are also available that provide alternatives to flush toilet, and sewer or septic tank technology.

Showers

People taking showers consume 18% of water in US homes[100] and about 20% of water in Australian homes.[101] A standard shower rose allows a water flow of about 8 litres per minute, so a 5-minute shower requires 40 litres of water. Water saving showerheads can cut consumption to less than one litre per minute, saving 35 litres of water and reducing the energy bill. Why the energy bill? Because hot water heaters only have to re-heat one-eighth the amount of water. If we use electric or gas hot water, using less water lowers energy bills and a home's contribution to greenhouse gas emissions.

Watering gardens

In some countries up to 50% of the water provided for domestic use is used to water gardens. This does not mean that all of the water pumped onto our vegetable gardens and roses is available to the plants. Watering the garden with a hose can be a relaxing, even meditative activity, but much of the water sprayed into the air evaporates before or as it hits the ground. The most efficient method is using drip or sub-surface root irrigation systems regulated with a timer. Those of us who rely on watering to keep us sane can pretend to squirt our plants by turning on our watering system, and standing in the garden with a length of disconnected hose!

Leaks

Drip, drip, drip. Leaking taps, toilets and showers are annoying and wasteful. One US study found that household leaks such as these made up one-tenth of the water consumption of a typical household. The worst offenders however, are the thousands of kilometres of water mains that bring water to our buildings. Bombay, India, loses a third of its water from leaking mains, US cities average losses of about a quarter of their total supply, while Manila in the Phillippines loses half due to leaks.[102]

The water in most cities is sourced remotely, transported through extensive networks of channels and pipes and, after use, having been contaminated is disposed of to ecosystems. Clean water is an increasingly scarce resource yet many buildings force people to use fresh water only once.

We can see from this discussion on stocks and flows that urban metabolisms are by and large a resource-hungry and polluting system. The accumulative result of the predominant use of natural systems as a source of resources and a sink for wastes is causing impacts that are altering geochemical cycles and threatening the health of ecosystems. BEEs need to understand the impacts of current practice in order to choose appropriate courses of action leading to ecological sustainability. These impacts are discussed in more detail in the next chapter.

References

1 Newton, P., Flood, J., Berry, M., Bhatia, K., Brown, S., Cabelli, A., *et al.* (1998) Environmental Indicators for National State of the Environment Reporting – Human Settlements. *Australia: State of the Environment Environmental Indicator Report.* Department of the Environment, Canberra, Australia.

2 Yencken, D. & Wilkinson, D. (2000) *Resetting the Compass: Australia's journey towards sustainability.* CSIRO Publishing, Melbourne.

3 Rees, W., Testemale, P. and Wackernagel, M. (1996) *Our Ecological Footprint: reducing human impact on the Earth.* New Society Publishers, Gabriola Island, Canada.

4 Howard, N. (2000) *Sustainable Construction: the data.* CR 258/99. Building Research Establishment Centre for Sustainable Construction Watford, UK.

5 Commonwealth of Australia (1995) *Greenhouse 21C: a plan of action for a sustainable future.* Department of the Environment, Sport and Territories, March, Canberra.   http://www.environment.gov.au/portfolio/esd/climate/air/climate/greenhouse/grn_21c.html.

6 World Resources Institute (1996) *World Resources 1996–1997: a guide to the global environment.* Oxford University Press, New York.

7 Newton, P.W., Brotchie, J.F. & Gipps, P.G. (1997) Cities in transition: changing economic and technological processes and Australia's settlement system. In: *Australia: state of the environment technical paper series (human settlements).* Department of the Environment, Canberra, Australia.

8 Augenbroe, G. & Pierce, A. (1998) *Sustainable Construction in the United States of America: a perspective to the year 2010.* CIB – W82 Report, June. Georgia Institute of Technology, Georgia, USA.

9 Yencken, D. & Wilkinson, D. (2000) *Resetting the Compass: Australia's journey towards sustainability.* CSIRO Publishing, Melbourne. pp.134–135.

10 Howard, N. (2000) *Sustainable Construction: the data.* CR 258/99. Building Research Establishment Centre for Sustainable Construction, Watford, UK. p.26.

11 Commonwealth of New Zealand (1997) Production and Consumption Patterns. Chapter 3. *State of New Zealand's Environment 1997.* Ministry for the Environment and GP Publications. New Zealand.

12 Nathan, D. (1999) Sending Rubbish Overseas. *The Straits Times.* Singapore, 08/04/99.

13 Steuteville, R. (1995) The state of garbage in America. *BioCycle* **6** (4) In: Peng, C., Scorpio, D. & Kibert, C. (1997) Strategies for successful construction and demolition waste recycling operations. *Construction Management and Economics* **15**, 49–58.

14 EcoRecycle Victoria (1999) Baulderstone Hornibrook. *Waste Minimisation Best Practice Case Study.* EcoRecycle Victoria, Melbourne, Australia.

15 Master Builders Association of Victoria, EcoRecycle Victoria, Royal Melbourne Institute of Technology (1998) *The Resource Efficient Builder* EcoRecycle Victoria, Melbourne, Australia. Vol.1 No.3 May.

16 Trusty, W. & Meil, J. (2000) The environmental implications of building new versus renovating an existing structure. In: Boonstra, C., Rovers, R. & Pauwels, S. (eds) *International conference sustainable building 2000 proceedings.* 22–25 October, Maastricht, Netherlands, Aeneas Technical Publishers. pp.104–106.

17 Housing Development Board (1999) *Annual Report.* HDB Singapore.

18 Roodman, D. & Lenssen, N. (1995) *A Building Revolution: how ecology and health concerns are transforming construction.* World Watch Paper No. 124, March. World Watch Institute, Washington DC, USA.

19 Jonsson, A. (2000) Industrial data on the material flows of the Swedish building stock. In: Boonstra, C., Rovers, R. & Pauwels, S. (eds) *International conference sustainable building 2000 proceedings.* 22–25 October, Maastricht, Netherlands, Aeneas Technical Publishers .p.305.

20 Shimoda, Y. & Mizuno, M. (2000) Material and energy metabolism in urban areas. In: Boonstra, C., Rovers, R. & Pauwels, S. (eds) *International conference sustainable building 2000 proceedings.* 22–25 October, Maastricht, Netherlands, Aeneas Technical Publishers. p.319.

21 Kibert, C.J. (2000) Deconstruction as an essential component of sustainable construction. In: Boonstra, C., Rovers, R. & Pauwels, S. (eds) *International conference sustainable building 2000 proceedings.* 22–25 October, Maastricht, Netherlands, Aeneas Technical Publishers. p.89.

22 Commonwealth of Australia (1996) *State of the environment Australia executive summary.* CSIRO Publishing, Melbourne, Australia.

23 World Resources Institute (1996) *World Resources 1996–1997: a guide to the global environment.* Oxford University Press, New York.

24 Hawken, P., Lovins, A., Lovins, L.H. (1999) *Natural Capitalism: creating the next industrial revolution.* Little, Brown & Co., New York.

25 Newell, B., Hall, D. & Molloy, R. (1999) *Port Phillip Bay Environmental Study Technical Reports.* CSIRO Environmental Projects Office, CSIRO Publishing: Melbourne.

26 Department of Primary Industries (1995) Urban Run-off Pollutes. *The Source* April, State Government of Queensland, Brisbane, Australia. In: Legge-Wilkinson (1996) *Human Impact on Australia's Beaches.* Wet Paper Publishing, Ashmore, Queensland, Australia.

27 Legge-Wilkinson (1996) *Human Impact on Australia's Beaches*. Wet Paper Publishing, Ashmore, Queensland, Australia. p.32.

28 Public Utilities Board Singapore (2002) www.pub.gov.sg. Accessed 11/2/02.

29 Newman, P. & Kenworthy, J. (1999) *Sustainability and Cities: overcoming automobile dependence*. Island Press, New York. p.76.

30 Critchley, B. (2000) Building material transport: UK practice. In: Boonstra, C., Rovers, R. & Pauwels, S. (eds) *International conference sustainable building 2000 proceedings*. 22–25 October, Maastricht, Netherlands, Aeneas Technical Publishers. p.192.

31 Department of Environment, Transport and the Regions (1997) Transport Statistics Great Britain. Department of Environment, Transport and the Regions, London, UK.

32 Howard, N. (2000) *Sustainable Construction: the data*. CR 258/99. Building Research Establishment Centre for Sustainable Construction, Watford, UK.

33 Newton, P.W., Brotchie, J.F. & Gipps, P.G. (1997) Cities in transition: changing economic and technological processes and Australia's settlement system. In: *Australia: state of the environment technical paper series (human settlements)*. Department of the Environment, Canberra.

34 Newton, P.W. (2000) Urban form and sustainable cities. In: Yencken, D. & Wilkinson, D. (2000) *Resetting the Compass: Australia's journey towards sustainability*. CSIRO Publishing, Melbourne. p.141.

35 Newton, P.W. (2000) Urban form and sustainable cities. In: Yencken, D. & Wilkinson, D. (2000) *Resetting the Compass: Australia's journey towards sustainability*. CSIRO Publishing, Melbourne.

36 O'Meara, M. (1999) *Reinventing cities for people and the planet*. World Watch Paper No. 147, June. World Watch Institute, Washington DC.

37 Commonwealth of New Zealand (1997) Production and Consumption Patterns. Chapter 2. *State of New Zealand's Environment 1997*. Ministry for the Environment and GP Publications.

38 Owen, K. (1996) Estate dreams end in despair. *Herald-Sun Newspaper* Melbourne, Australia. 22/01/96.

39 Thomas, M. & Edwards, S. (2000) Transport and buildings: reducing the environmental impact. In: Boonstra, C., Rovers, R. & Pauwels, S. (eds) *International conference sustainable building 2000 proceedings*. 22–25 October, Maastricht, Netherlands, Aeneas Technical Publishers.

40 Yates, A., Baldwin, R., Howard, N. and Susheel Rao (1998) *BREEAM 98 for Offices*. BRE Report 350, Building Research Establishment, Watford, UK.

41 Scheurer, J. (2001) Residential Areas for Households without Cars. Paper presented at *Trafikdage pa Aalbourg Universitet* 27–28 August, Sweden.

42 Kenworthy, J. (1998) City building and transportation around the world. In: *Village wisdom future cities – proceedings of the third international ecocity and ecovillage conference*. 8–12 January 1996, Yoff, Senegal, Ecocity Builders, Oakland USA. p.32.

43 Thomas, M. & Edwards, S. (2000) Transport and buildings: reducing the environmental impact. In: Boonstra, C., Rovers, R. & Pauwels, S. (eds) *International conference sustainable building 2000 proceedings*. 22–25 October, Maastricht, Netherlands, Aeneas Technical Publishers.

44 President's Council on Sustainable Development (PCSD) (1996) Vision Community Capacity Building. *PCSD Sustainable Communities* Chapter 2. The White House, Washington DC, USA.

45 Australian Bureau of Statistics (1996) *Environmental Issues: people's views and practices.* ABS Catalogue No. 4602.0, February. Canberra.

46 Augenbroe, G. & Pierce, A. (1998) *Sustainable Construction in the United States of America: a perspective to the year 2010.* CIB – W82 Report, June. Georgia Institute of Technology, Georgia, USA.

47 Commonwealth of Australia (1996) *State of the environment Australia executive summary.* CSIRO Publishing, Melbourne, Australia.

48 Kong, L., Yuen, B., Sodhi, N. & Briffett, C. (1999) The construction and experience of nature: perspectives of urban youths. *Tijdschrift voor Economische en Sociale Geografie* **90** (1) 3–16.

49 American Association of Engineering Societies and World Engineering Partnership for Sustainable Development (1996) *The Role of Engineering in Sustainable Development.* American Association of Engineering Societies, Washington DC, USA.

50 Royal Australian Institute of Architects (1995) RAIA Environment Policy. *BDP Environment Design Guide* General Issues Paper 1, May, Melbourne, Australia.

51 Baulderstone Hornibrook Bilfinger + Berger (BHBB) (2000) *M5 East Motorway: environmental management overview.* BHBB M5 East Joint Venture, Issue No. 1, September. Sydney, Australia.

52 Odum, H.T. & Odum, E. (1976) *Energy Basis for Man and Nature.* McGraw Hill, New York.

53 Bush, S., Holmes, L. & Luan, H.T. (1995) *Australian Energy Consumption and Production: historical trends and projections to 2009–10.* Australian Bureau of Agricultural and Resource Economics. Research Report No. 95.1. Canberra, Australia.

54 Tucker, S.N., Salomonsson, G., Treloar, G., MacSporran, C. & Flood, J. (1993) *The Environmental Impact of Energy Embodied in Construction.* Main Report for the Research Institute of Innovative Technology for the Earth. CSIRO Division of Building, Constructon and Engineering. Highett, Australia.

55 Commonwealth of Australia (AGO) (1999) *Australian commercial building sector greenhouse gas emissions 1990–1910, executive summary report.* Australian Greenhouse Office, Canberra, Australia. p.13.

56 Treloar, G.J. (1996) *The Environmental Impact of Construction: a case study.* Australia and New Zealand Architectural Science Association, Sydney, Ausralia. p.49.

57 Tucker, S.N., Salomonson, G.D., Ambrose, M.D., Treloar, G.J., Hunter, B., Edwards, P.J., et al. (1996) *Development of Analytical Models for Evaluating Energy Embodied in Construction.* Report to the Energy Research and Development Corporation, CSIRO Division of Building, Construction and Engineering, May, Highett, Australia.

58 Lawson, B. (1995) Embodied energy of building materials. *Environment Design Guide.* Royal Australian Institute of Architects. PRO 2, August, 3.

59 Miller, A. (1997) Transportation Energy Embodied in Construction Materials. *Second International Conference – Buildings and the Environment.* 9–12 June, Centre Scientifique et Technique du Batiment, Paris, France. pp.477–484.

60 Critchley, B. (2000) Building material transport: UK practice. In: Boonstra, C., Rovers, R. & Pauwels, S. (eds) *International conference sustainable building 2000 proceedings.* 22–25 October, Maastricht, Netherlands, Aeneas technical publishers.

61 Critchley, B. (2000) Building material transport: UK practice. In: Boonstra, C., Rovers, R. & Pauwels, S. (eds) *International conference sustainable building 2000 proceedings.* 22–25 October, Maastricht, Netherlands, Aeneas technical publishers.

62 Augenbroe, G. & Pearce, A. (1998) *Sustainable Construction in the United States of America: a perspective to the year 2010.* CIB – W82 Report, June. Georgia Institute of Technology, Georgia, USA.

63 Fay, R. & Treloar, G. (1998) The embodied energy of living. In: *The Environment Design Guide.* General Issues paper 20, August. The Royal Australian Institute of Architects, Melbourne, Australia.

64 Kibert, C.J. (1994) Establishing principles and a model for sustainable construction. In: *Sustainable construction, proceedings of the first international conference of TG16.* 6–9 November, Tampa, Florida, USA.

65 Rees, W. (1999) The built environment and the ecosphere: a global perspective. *Building Research and Information* **27** (4/5) 206–220.

66 Howard, N. (2000) *Sustainable Construction: the data.* CR 258/99. Building Research Establishment Centre for Sustainable Construction, Watford, UK.

67 Roodman, D. & Lenssen, N. (1995) *A Building Revolution: how ecology and health concerns are transforming construction.* World Watch Paper No. 124, March. World Watch Institute, Washington DC, USA.

68 Howard, N. (2000) *Sustainable Construction: the data.* CR 258/99. Building Research Establishment Centre for Sustainable Construction, Watford, UK. p.10.

69 Rees, W., Testemale, P. & Wackernagel, M. (1996) *Our Ecological Footprint: reducing human impact on the Earth.* New Society Publishers, Gabriola Island, Canada.

70 Rees, W. (1999) The built environment and the ecosphere: a global perspective. *Building Research and Information* **27** (4/5) 206–220.

71 Rees, W., Testemale, P. & Wackernagel, M. (1996) *Our Ecological Footprint: reducing human impact on the Earth.* New Society Publishers, Gabriola Island, Canada.

72 Weizsäcker, E.von, Lovins, A.B. & Lovins, L.H. (1997) *Factor four: doubling wealth - halving resource use: the new report of the Club of Rome.* Allen & Unwin, Sydney, Australia.

73 Sampat, P. (2000) Groundwater shock: the polluting of the world's major freshwater stores. *World Watch.* World Watch Institute, Washington DC, USA. **13** (1), 10–22.

74 Flavin, C. (1995) Natural gas production stalls. In: World Watch Institute *Vital Signs 1995-1996.* Earthscan Publications, London.

75 Flavin, C. (1995) Natural gas production stalls. In: World Watch Institute *Vital Signs 1995-1996.* Earthscan Publications, London.

76 Carley, M. & Christie, I. (1993) *Managing Sustainable Development*. The University of Minnesota Press, Minneapolis, USA.

77 Flavin, C. (1995) Natural gas production stalls. In: World Watch Institute *Vital Signs 1995-1996*. Earthscan Publications, London.

78 World Resources Institute (1994) *World Resources 1994–1995: a guide to the global environment*. Oxford University Press. New York.

79 Crowther, P. (2000) Building Deconstruction in Australia. *International Status of Deconstruction* CIB TG39. In: Kibert, C.J. (2000) Deconstruction as an essential component of sustainable construction. In: Boonstra, C., Rovers, R. & Pauwels, S. (eds) *International conference sustainable building 2000 proceedings*. 22–25 October, Maastricht, Netherlands, Aeneas Technical Publishers. p.90.

80 The One Stop Timber Shop. www.timbershop.wilderness.org.au

81 Kibert, C.J. (2000) Deconstruction as an essential component of sustainable construction. In: Boonstra, C., Rovers, R. & Pauwels, S. (eds) *International conference sustainable building 2000 proceedings*. 22–25 October, Maastricht, Netherlands, Aeneas Technical Publishers. p.89.

82 Golton, B. (1994) Affluence and the ecological footprint of dwelling in time: a Cyprus perspective. In: *First world conference on sustainable construction CIB TG16*. In: Smith, M., Whitelegg, J. & Williams, N. (1999) *Greening the built environment*. Earthscan Publications, London. p.64.

83 Gelder, J. (1996) A Survey of Chemical Risk in the Built Environment. In: *The Environment Design Guide*. Product Issues paper 5, February. The Royal Australian Institute of Architects, Melbourne, Australia.

84 Schmidt-Bleek, F. (1994) Carnoules declaration of the Factor Ten Club. Wuppertal Institute. In: Weizsäcker, E. von, Lovins, A.B. & Lovins, L.H. (1997) *Factor four: doubling wealth: halving resource use: the new report of the Club of Rome*. Allen & Unwin, Sydney, Australia.

85 Crowther, P. (2000) Building Deconstruction in Australia. *International Status of Deconstruction* CIB TG39. In: Kibert, C.J. (2000) Deconstruction as an essential component of sustainable construction. In: Boonstra, C., Rovers, R. & Pauwels, S. (eds) *International conference sustainable building 2000 proceedings*. 22–25 October, Maastricht, Netherlands, Aeneas Technical Publishers.

86 Ministry of Environment Singapore (2000) *Waste Minimisation Data and Statistics 2000* at: http://www.env.gov.sg/info/WASTE/Data%20and%20Statistics.htm#2000. Accessed 6/02/02.

87 Kibert, C.J. & Chini, A. (eds) (2000) *Overview of deconstruction in selected countries, CIB Report, TG39*. CIB, International Council for Research and Innovation in Building, Construction Task Group 39 CIB.

88 Howard, N. (2000) *Sustainable Construction: the data*. CR 258/99. Building Research Establishment Centre for Sustainable Construction, Watford, UK. p.23.

89 ANZECC (1997) *Wastewise Construction Review Report – Draft*. Australian and New Zealand Environment Conservation Council, 2 September, Sydney NSW.

90 Core C&D Waste. Estimates from Member States, OECD and Study Team in Symonds Group Ltd in association with ARGUS, Cowl and PRC Bouwcen-

trum (1999) *Construction and demolition waste management practices and their economic impacts, Report to DGXI, European Commission, Final Report.* February.

91 Hong Kong Polytechnic (1993) *Reduction of construction waste: final report.* The Hong Kong Construction Association Ltd, Hong Kong.

92 Franklin Associates (1998) *Characterisation of Building-Related Construction and Demolition Debris in the United States.* USEPA Report No. EPA530-R-98-010, US EPA Municipal and Industrial Solid Waste Division, Los Angeles, USA. June.

93 Gavilan, R.M. & Bernold, L. (1993) Evaluation of Solid Waste in Building and Construction. *Draft report for ASCE Journal of Construction Engineering and Management.* American Society of Engineers, Reston, VA, USA. June.

94 Formoso, C., Frianchi, C. & Soibelman, L. (1993) Developing a Method for Controlling Material Waste on Building Sites. *Proceedings of Economic Evaluation and the Built Environment.* Lisbon, CIB Rotterdam, Netherlands.

95 Graham, P. & Smithers, G. (1996) *Construction waste minimisation for Australian residential development Asia Pacific Journal of Building and Construction Management.* **2** (1) 14–18.

96 Hong Kong Polytechnic (1993) *Reduction of construction waste: final report.* The Hong Kong Construction Association Ltd, Hong Kong.

97 US Department of Commerce (2000) *National Trade Data Bank*, published November 3, 2000 at http://www.tradeport.org/ts/countries/singapore/isa/isar0022.html. Accessed 11/02/02.

98 Hawken, P., Lovins, A. & Lovins, L. H. (1999) *Natural Capitalism: creating the next industrial revolution.* Little, Brown & Co. New York, Chapter 11.

99 Hawken, P., Lovins, A. & Lovins, L. H. (1999) *Natural Capitalism: creating the next industrial revolution.* Little, Brown & Co. New York, Chapter 11.

100 Hawken, P., Lovins, A. & Lovins, L. H. (1999) *Natural Capitalism: creating the next industrial revolution.* Little, Brown & Co. New York, Chapter 11.

101 Waldock, T. (1996) An Introduction to Water Sensitive Design. *BDP Environment Design Guide* Design Paper 13, November. Melbourne, Australia.

102 US Water News (1992) Leakage Varies World Wide. April. p.18. In: Hawken, P., Lovins, A. & Lovins, L. H. (1999) *Natural Capitalism: creating the next industrial revolution.* Little, Brown & Co. New York, Chapter 11.

4 IMPACTS: THE EFFECTS OF CURRENT PRACTICE

Introduction

'Broadly stated, the common pattern in the deterioration of organic life as a result of building intrusion (both in terrestrial and aquatic ecosystems) includes a decline in the biomass (the mass of living matter), a decline in productivity (the amount of material produced by a given species present in a given area), and the malfunctioning of natural controls.'[1]

The type and quantity of resource input flows and the volume and harmfulness of the output flows of a building development, over its lifetime, contribute to its likely impact on ecosystems and people. BEEs know that if resources flowing into a building development can be stored and drawn upon as stocks of resources, then the damage to ecosystems as a result of a building's flows can be reduced. Impacts occur at and around the development site, within the bioregion and in remote ecosystems. The impacts at each scale are interrelated, which means every local action can have global consequences.

Impact categories

Often the immediate impacts caused by construction are on surrounding built environments. Natural environments remote from the construction project are impacted upon at the same time due to the material supply chain but it is often difficult to identify the direct impacts caused by a particular project. We might be expecting a lot from our BEEs if we expected them to remember every individual impact that the metabolism of a building or built environment might cause. There are more impacts than we could possibly know. Building projects may impact on natural environments that are far removed from the site and may be accumulative and long-term.

So, rather than listing all of the impacts highlighted in the previous section, adding to that list all of the impacts *not* discussed, and having a test, we will do what BEEs do, and think in terms of the categories of impact that our building activity can cause. Ramachandran[2] identified four categories of environmental impact associated with building. They are:

- resource depletion;
- physical disruption;
- pollution; and
- social and cultural effects.

Resource depletion

The first relates to impacts caused by the use of natural resources in the creation and operation of buildings. Resources are generally viewed as primary materials such as fossil fuels, mineral ores and timber. Of primary concern is our use of non-renewable resources because we cannot replenish the stores of material we use. However, it is possible to deplete renewable resources also. Availability of renewable resources such as timber is diminishing in some countries through poor management of the extraction process.[3]

BEEs don't only consider primary materials as resources. They also consider the ecological systems and functions that create primary materials as resources. The effects of building activity on ecological systems and services such as native forests, rivers, wetlands, clean air and clean water are also considered during development decision-making. The ability of these systems' resources to function in ways that provide the materials and services we need should also be considered as resources depleted by building development, even if the primary materials being extracted for physical use in construction or operation of buildings are renewable.

Timber, for example, is renewable because it can be regrown, but removing it from a rainforest may damage the integrity of the ecosystem and therefore deplete not only trees but forest resources and services. Rate of regeneration is also an issue. Native hardwoods, for example, can be regrown, but the forests they come from are complex ecosystems that cannot be easily or quickly replaced.

Physical disruption

Physical disruption relates to the interference caused to the environment by the actual construction process, and again when the building is occupied. The physical disruption to the surrounding environment could also be caused by maintenance, refurbishment or demolition works, as well as by the operations carried out within the facility during its lifetime.

Yencken and Wilkinson[4] identify three major environmental issues associated with physical disruption caused by the construction process.

- *Loss of productive land*: where construction sites have not been previously built on. The capacity of a site to provide habitat or land for agricultural production needs to be considered. Building sites that are not ecologically or agriculturally productive, normally because they have been previously built on, should be selected.
- *Disturbance due to building development* such as soil erosion and the dispersal of sediments and pollution into waterways: where the likelihood of soil erosion and water pollution due to sediment run-off is heightened during excavation works.
- *Degradation and loss of biodiversity* in surrounding hinterlands: due to material choice where material extraction damages ecosystems. This might be caused by the purchasing of native forest timber, mined materials with large ecological rucksacks, or use of materials produced by polluting manufacturing processes.

Physical disruption may be caused equally to a surrounding built environment. Effects of noise, dust or wastewater run-off from building sites may be just as disruptive in an urban environment as they would be in a natural environment. Physical disruption may also refer to noise pollution or vibration caused by the construction process[5] or by the operation of a building during its life cycle.

Pollution

Pollution caused by the production of materials and the construction and operation of buildings impacts on the natural and built environment. The effects of building-related pollution could also contribute to resource depletion if, for example, pollution occurs in a water source, or contaminates land.

Building in recent times has had almost as equally damaging effects on people as on ecosystems. The increased use of synthetic materials, mechanical control of ventilation, coupled with poor consideration of the relationships between material choice, climate, site and their combined effect on people, has led to the creation of unhealthy indoor environments. Human health and comfort are fundamental requirements for sustainable buildings. Pollution of the indoor environment presents direct health risks to people. Many common materials contain substances that are toxic and which can lead to serious health impacts. Poor indoor environmental quality affects people's working environment, leading to a decrease in productivity if workers are made ill through exposure to fumes from off-gassing materials, bacteria or viruses in poorly maintained ventilation systems, or lack of fresh air. Buildings that make people sick are called sick buildings. Being made ill by a building is generally called sick building syndrome.

Sick building syndrome is a phenomenon that has motivated building designers to consider the toxicity of many common building materials and, in particular, interior finishes. The off-gassing of materials, particularly those containing formaldehydes and xylene, are commonly incorporated as environmental impact criteria for building materials.[6]

Australians are suffering from the respiratory illness, asthma, in increasing numbers. One of the main triggers of asthma attacks is inhalation of microscopic faeces from the house dust mite, which lives in carpets, bedding, curtains, and places where dust gathers such as the tops of pelmets and architraves. The design and specification of interior finishes and furnishing therefore has a significant influence on asthma sufferers.

Another aspect of pollution, particularly from synthetic waste, is its resilience. Land can remain contaminated for years after being polluted by chemicals and this can pose health risks for construction workers or people inhabiting buildings constructed on contaminated land.

Social and cultural effects

The final issue identified by Ramachandran deals with effects in the social and cultural environment which may not be as easily detected as resource depletion, physical disruption or pollution, particularly in already-built environments. The social environment in an area may suffer through development. The development of a grand prix racetrack on a recreational reserve would cause a large social impact to a residential environment even if only used sporadically. There could of course be positive impacts to a social or cultural environment as a result of a development such as a community or cultural centre.

The cultural sensitivities of people affected by a development must also be considered. The impact of building development in historical districts or land under the custodianship of indigenous people, for example, can require extensive public consultation and careful planning so that cultural values are appropriately identified and catered for. Social and cultural issues and impacts are far harder to identify and, in many cases mitigate, than scientifically determinate issues such as energy consumption or material toxicity. It is essential that they be addressed.

Buildings also cause a visual impact on the environment. Whether visual impact is positive or negative may be a question of individual taste, nevertheless the aesthetics of a development do impact on the surrounding environment and are perhaps one of the most noticeable of all the environmental impacts of building. There is little doubt that people affected by a building development will offer more opinion on its visual impact than they would about the level of resource depletion caused by its construction or operation.

BEEs use impact categories such as resource depletion, physical disruption, pollution and social and cultural effects as a mental checklist when they are making project decisions. They always try to make decisions that will minimise both the likelihood and magnitude of the impacts described.

Impacts on global cycles

The term 'global cycles' refers to *biogeochemical cycles*: the planetary processes that cycle matter through the biosphere. The accumulating environmental impacts of human industry has, predominantly since the industrial revolution, destabilised these flows, leading to global changes in climate and ecological health that are beginning to present limits to the stability and growth of human economies.

In 1972 the scientific organisation, the Club of Rome, commissioned a report to determine whether, given the predominant reliance of the industrialised economy on finite resources, there would ever be a limit to its ability to meet human needs. The report, entitled *Limits to Growth*, found that it was possible that diminishing resources and accumulating pollution would limit economic growth. The report concluded:

'If the present growth trends in world population, industrialisation, pollution, food production, and resource depletion
continue unchanged, the limits to growth on this planet will be
reached sometime in the next one hundred years'.[7]

There is no doubt that the human race's technological development
has been staggering in recent times. The rate of development has also
been steadily increasing to the point where technologies are becoming obsolete well before the materials they are made of, and an entire
infrastructure for dealing with waste from a 'throw-away' society
has developed.

The findings of *Limits to Growth* spurred environmental reforms
in many countries, but its prediction that economic growth would
be limited by patterns of resource use has not yet come to pass. Some
researchers consider that this might be explained by increasing industrial resource efficiency or our increasing ability to extract new
resources in economical ways.[8] Yet, seemingly unconstrained, global
resource consumption continues to increase.[9]

While the *Limits to Growth* perspective argues that reduced resource availability is a constraining factor, it now seems that it poses
less of a threat to human economies than the side effects of our resource consumption patterns. As R. Buckminster-Fuller once wrote:
'...what humanity rated as "side-effects" are nature's main effects'.[10]
We are facing limits to continued industrial expansion, but rather
than focusing on when we might run out of a resource, our attention
has turned now to the impacts on Earth's life-supporting global biogeochemical cycles, of the pollution and waste associated with our
consumption patterns.

Global cycles – what BEEs know

Ecosystems are driven by 'global cycles', complex movements and
transfers of matter and energy from the atmosphere, through the
ecosphere (plants, animals and other life forms – both dead and alive)
to the lithosphere (Earth's crust) and back again.

These 'natural cycles' are known scientifically as biogeochemical
cycles and are thought to have been cycling chemical elements through
ecosystems in relatively stable flows until the beginning of the industrial revolution. The five major biogeochemical cycles upon which life
depends are the carbon cycle, the hydrological (water) cycle, the nitrogen cycle, the phosphorus cycle, and the sulphur cycle. The elements in
these cycles are stored at different stages through their cycle. Carbon,

for example, is stored in the Earth's crust as carbonate and hydrocar-bons, in plants and animals, and in the atmosphere as oxides.

While each element cycles relatively quickly through the eco-sphere and atmosphere (with the exception of phosphorus, which is not present in the atmosphere), most elements cycle slowly through the lithosphere and are released equally slowly back to ecosphere and atmosphere. This means that over many millennia there has been an accumulation of important substances like fossil fuels stored in the Earth's crust. We have, however, in little more than 200 years so rapidly increased the release of elements from the Earth's crust back into the *ecosphere* and atmosphere that the biogeochemical cycles regulating Earth's life-support systems have been destabilised. The accumulation of industrial pollution, manifesting as effects such as global warming, declining ecosystem health, acid rain and ozone depletion are now considered major threats to our capacity to follow industrialised patterns of development.

Disruption of natural cycles has led to increased carbon, nitrous (NO_x) and sulphurous oxides (SO_x) in the atmosphere, contributing to global warming and acid rain respectively. Phosphorus levels have been increased in soils and water leading to increased nutrient levels in aquatic ecosystems. In some cases this has led to algal blooms and the strangling of waterways by overgrowth of aquatic plants.

Traditional building practice uses matter and energy to produce buildings that consume yet more resources, and at each stage of the building life cycle waste is accumulated. The UK construction industry, for example, is responsible for approximately 2.5% of NO_x emissions and 8% of SO_x emissions nationally. In addition the trans-portation of construction materials in the UK contributes about 10% of national CO_2 emissions.[11] Very little of the surplus material from mining, material manufacture, building construction and operation is recycled in any area of our industrialised economy, let alone used as inputs to create new buildings.

BEEs take action to help restabilise biogeochemical flows. Because we want to know what BEEs know, we need to examine the impacts of building and built environments in each of these cycles.

Building and the carbon cycle

> 'Geologists estimate that every carbon atom on Earth has made about 30 round trips over the past 4 billion years.'[12]

Carbon atoms are essential building blocks of life as we know it. For organisms they serve three main purposes.

Building's major influences on the carbon cycle

Fossil fuel consumption in

- Material production
- Transportation
- Generating energy to run buildings

Major impacts

- Greenhouse gas emission and contribution to climate change

- They are the structural components of organic molecules;
- The chemical bonds carbon forms to store energy; and
- The atmospheric form of CO_2 traps reflected long-wave heat radiation near the Earth's surface, causing a 'greenhouse effect' that has kept Earth's surface temperature within a range that sustains life.

Carbon exists in the atmosphere and is absorbed into the ecosphere by terrestrial and aquatic plants during photosynthesis, through which it is transformed (with oxygen) into sugars. The energy source for this chemical reaction is solar radiation. Carbon-absorbing elements of the carbon cycle, such as plants, are called 'carbon sinks'.

From its storage in plants some is released again into the atmosphere during respiration, and some is stored in plant cells. Carbon stored in a plant cell is released when the plant dies and begins to decay or is burnt. In a similar way, carbon absorbed by animals such as humans (when we eat sugars) may be used for respiration, in which case carbon (in the form of CO_2) is quickly released into the atmosphere as we breathe out. Carbon used by our bodies as part of our cellular structure will remain in our bodies until we eventually decay or are burnt.

Huge deposits of carbon from decaying organic matter have become compressed over time by geological processes and have become hydrocarbons (fossil fuels). Carbon in this state has been stored for millions of years, as has the carbon that has combined with calcium in shells and bones of aquatic animals and become limestone and dolomite ($CaCO_3$). These limestone deposits have locked carbon away from the atmosphere for very long periods, slowly releasing carbon through weathering. The carbon cycle supports life by regulating the quantity of carbon in the atmosphere, thus contributing principally to favourable atmospheric and weather conditions, and contributing to the respiration of plants which in turn releases oxygen that animals need to breathe. So where does human activity and building in particular fit into the carbon cycle?

A more rapid release of carbon into the atmosphere in the form of CO_2 has occurred since the industrial revolution due principally to the burning of fossil fuels. This has in turn caused a destabilisation of the carbon cycle leading to increasing concentrations of atmospheric CO_2 and an increase in global mean temperature, a phenomenon for which there is significant consensus of scientific opinion.[13] By 2100 atmospheric CO_2 levels are projected to be up to 250% greater than they were in 1750.[14]

Modern building is both created and sustained by industrial activity. The building process is one of the major contributors to the release of CO_2 and other so-called *greenhouse gases* from the lithosphere to the atmosphere. The major impacts from the building industry come from fossil fuel consumption in building material manufacture, for transportation, and in providing electricity for the operation of buildings.

Material production

In relation to the carbon cycle, the building industry is responsible for emissions when fossil fuel energy (primary energy such as coal, oil or natural gas) is consumed to create electricity or fuel (delivered energy) for the mining and manufacture of building materials, and for their transportation. The manufacturing processes of many common building materials also cause the release of CO_2 due to the burning of coal in the production process. Steel manufacture, for example, is estimated to cause the emission of approximately two tonnes of CO_2 for every one tonne of steel produced,[15] while cement manufacture causes approximately one tonne of CO_2 per tonne of cement.[16]

The embodied energy of a material can be used as an indicator of its contribution of CO_2 equivalent greenhouse gas emissions. However, accurately determining this depends on how accurately the CO_2 emissions from different fuels used to manufacture the product is known (see Table 4.1).

An indication of the disruptive influence of a material on the carbon cycle can then be gained by relating the embodied energy of a material with the actual quantities produced by the construction industry globally. To explain this we will use the manufacture of Portland cement, the world's most used building material, as an example.

Example – the effect of cement manufacture on the carbon cycle

Portland cement contains oxides of calcium, silicon, aluminium and iron. The primary raw material is limestone, which is quarried from deposits of varying concentrations. The amount of raw material mined to produce Portland cement therefore exceeds the actual quantity of the material produced. In the US during 1999, 86 megatonnes (Mt) of Portland cement was produced from 140 Mt of raw material.[17] The ecological rucksack of Portland cement can be up to two tonnes of raw material per tonne of cement[18] although some estimates put the figure as high as 10 tonnes mined per tonne of raw material produced.[19] Because of the fuel consumed in mining, this process inefficiency contributes greatly to the embodied energy of Portland cement.

Raw materials are crushed, blended and then calcinated in high temperature kilns run either with electricity or natural gas. There are two processes that produce CO_2 emissions at this stage. First, CO_2 is a natural byproduct of the calcination reactions that raw materials undergo. Second, the fuel used to fire the kilns causes CO_2 emissions. Cement production in the US involves, on average, 0.5 tonnes of CO_2 per tonne of cement being released during calcination, and an additional 0.44 to 0.5 tonnes per tonne of cement due to fuel consumed in kilns. Total emissions of CO_2 can therefore be averaged at approximately 1 tonne per tonne of cement.[20]

Global production of cement in 1999 was estimated at 1.6 billion tonnes, and in 2000 this had risen to 1.7 billion tonnes.[21,22] Contribution of CO_2 from cement manufacture is therefore approximately the same. Cement manufacture is estimated to be contributing about 4% of global CO_2 emissions annually (see Fig. 4.1).[23]

Transportation

As we discussed earlier, the transport of building materials and the commuting patterns of building occupants can have a significant influence on energy consumption. Carbon emissions from petroleum and diesel fuel consumption are an important consideration given that most movement of both building material and people in many countries is by road. The mode of transport chosen by tradespeople to get to and from construction projects is also a significant influence on building-related greenhouse gas emissions.

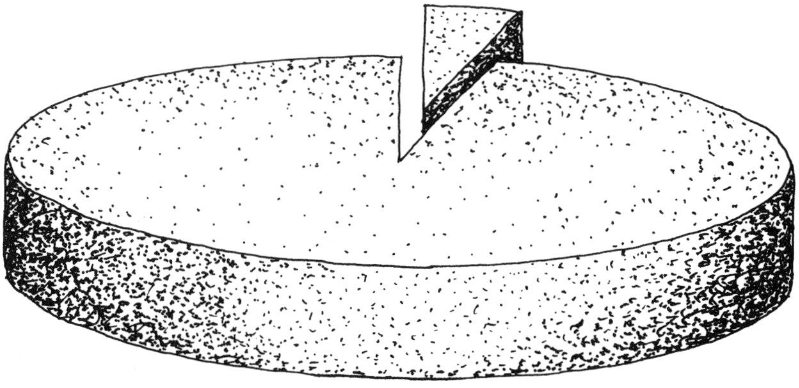

4% cement manufacture

Contribution of cement manufacture to total annual CO2 emissions

Fig. 4.1 Carbon dioxide emissions from cement manufacture as a percentage of global emissions. In the year 2000 global production of cement resulted in the emission of about 1.7 billion tonnes of carbon dioxide. Reference: Marland, G., Boden, T.A. & Andres, R.J. (2001) Global, Regional, and National Fossil Fuel CO_2 Emissions. In: *Trends: A Compendium of Data on Global Change.* Carbon Dioxide Information Analysis Center, Oak Ridge National Laboratory, US Department of Energy, Oak Ridge, TN, USA.

Energy to operate buildings

The major contribution of CO_2 to the atmosphere attributable to the activities of the building industry is in creating building systems that run on electricity generated from fossil fuels, or through the direct consumption of fuel oil or natural gas.

The quantity of CO_2 released over a building's life cycle depends on the fuels used to generate and supply its energy. Different fuels release different amounts of CO_2 when they are burned, as Table 4.1 shows.[24]

Buildings or building materials that derive large amounts of electrical energy generated from coal- or oil-fired power stations obviously contribute to the release of more CO_2 emissions than if the building materials or building derived energy from natural gas. Energy from non-carbon based sources such as hydro-electricity and nuclear power contributes little carbon to the atmosphere, however, both forms of energy production pose other potentially devastating (and more immediate) ecological threats. Renewable energy sources

Table 4.1 Carbon dioxide emissions for various fuels.[23]

Fuel	CO_2 (kg/GJ)
Natural gas	55.0
Petroleum products	77.0
Black coal	91.7
Brown coal	95.3
Electricity (coal fired)	286.0

Source: Ecologically Sustainable Development Working Groups (1991) *Final Report – Energy Use.* AGPS, Canberra, 168. In Bill Lawson (1996) *Building Materials Energy & the Environment: Towards Ecologically Sustainable Development.* Royal Australian Institute of Architects, Sydney, 12.

such as solar and wind energy are becoming viable low-impact solutions.

The World Watch Institute recently estimated that buildings consume about 40% of energy produced worldwide.[25] However, this quantity of energy consumed can be greater in more industrialised and urbanised countries. In the UK, for example, 56% of all energy produced is used to operate buildings; 10% of all energy produced is used to manufacture building materials. Therefore over two-thirds of all UK energy use stems from or is associated with building construction and use.[26]

Considering how much energy is actually consumed by building, BEEs appreciate that even if an effort is made to use energy sources with low associated CO_2 emissions, the sheer volume of fuel consumed implies that CO_2 emissions will continue to be released into our atmosphere at historically unprecedented rates. In Australia, for example, the contribution of greenhouse gases from the commercial building sector are expected to nearly double from the 1990 level of 32 Mt of CO_2 per annum to 63 Mt per annum by 2010 (see Fig. 4.2).[27] Greenhouse emissions attributable to heating and cooling Australian residential buildings are expected to increase by between approximately 14% and 39% in the same period.[28] This observation presents a more urgent imperative for BEEs, and that is to minimise life-cycle building energy consumption.

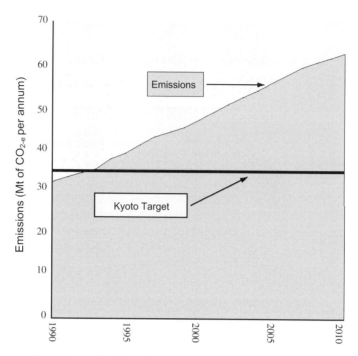

Fig. 4.2 Projections of carbon dioxide emissions from Australian commercial buildings. Research by the Australian Greenhouse Office indicates that greenhouse gas emissions associated with energy consumption in Australian commercial buildings will rise to 63 Mt per annum, almost double the 1990 level of 32 Mt per annum. Source: Commonwealth of Australia (AGO) (1999a) *Australian commercial building sector greenhouse gas emissions 1990–1910, executive summary report.* Australian Greenhouse Office, Canberra, Australia. p.9 Fig 1.

Building and the nitrogen cycle

Nitrogen is an essential element in our DNA and RNA and is there-

---

Building's major influences on the nitrogen cycle

- Fuel consumption for construction processes
- Cement manufacture

Major impacts

- Greenhouse emissions

---

fore integral to our genetic blueprint. Nitrogen oxides are greenhouse gases and are also very important fertilisers. Nitrogen (N_2) makes up 70% of the Earth's atmosphere and is essential to the creation of amino acids, proteins and peptides – the building blocks of life. The trouble is that plants and animals can't use nitrogen in its atmospheric form. Nitrogen is made useable by the armies of bacteria that move atmospheric nitrogen through the lithosphere into the ecosphere and back to the atmosphere again.

First nitrogen 'fixing' bacteria convert nitrogen to ammonia (NH_4), different bacteria convert ammonia to nitrite (NO_2^-), and yet more bacteria convert nitrites to nitrates (NO_3), a form which plants can absorb. Plants then turn nitrates into amino acids, proteins and peptides, which are in turn eaten by animals. Nitrogen is released into the environment again when animals die and decompose, and when other organic matter such as fallen leaves, flowers and hair decompose. It is also released by human activity, particularly through burning fossil fuels in vehicles; industrial processes like cement manufacture, and through the use of manufactured nitrate fertilisers. The extensive use of nitrate-based fertilisers has in some areas overloaded groundwater and watersheds. Poor agricultural practices are therefore major disruptive factors.[29]

Because of the increased global rate of industrialisation, fossil fuel consumption and fertiliser use, levels of nitrogen oxides (NO_x) have been accumulating in our atmosphere and some aquatic environments at rates far exceeding the rate at which the natural world's bacteria can 'fix' atmospheric nitrogen (a process known as 'denitrification'). This is causing a number of severe problems. The concentration of nitrous oxide (N_2O) is estimated to be increasing by about 0.2% annually (Fig. 4.3) and is thought to contribute about 6% of the global warming potential of greenhouse gases.[30,31] Nitrogen dioxide (NO_2) is the element in photochemical smog that gives it its orange tinge. When NO_x are diluted in water in the atmosphere they form nitric acid (HNO_3) and fall as acid rain. The building industry contributes to this increase in atmospheric NO_x principally through the manufacture of cement, and through the fossil fuel energy embodied in building materials, and energy used during construction.

Cement manufacture

The manufacture of cement is estimated to contribute between 1.8 and 6.0 kg of NO_x per tonne of cement; the variability in the quantity depends on the type of fuels used to calcinate limestone.[32] The sheer volume of building materials made with cement provides some indication of the amount of NO_x that the industry is responsible for

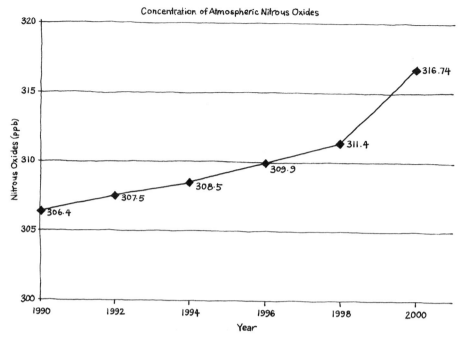

Fig. 4.3 Increasing atmospheric nitrogen. Measurements of levels of atmospheric nitrous oxides reveal an increase of approximately ten parts per billion (ppb) in ten years. The building industry contributes to this atmospheric loading of nitrous oxides principally through cement manufacture. Reference: World Resources Institute (2002) *Earth Trends the Environmental Information Portal*. Climate and atmosphere data tables. http://earthtrends.wri.org/searchable_db/index.cfm. Accessed 04/02/02.

worldwide. The European building industry consumes more than 175 million tonnes of cement each year.[33] If we adopted an average of 3.9 kg of NO_x released for every tonne of cement produced, based on the figures quoted earlier for global production of 1.7 billion tonnes, in the year 2000 the building industry contributed to the emission of about 6630 tonnes of NO_x to the atmosphere.

Atmospheric nitrogen loading can be reduced most obviously by using less material containing cement, and by using low-cement materials. The approach taken in ecologically sustainable building projects is therefore firstly to maximise the efficiency with which cement-rich materials such as concrete and mortar are used through good structural design. The other approach is to use pozzolanic (cement-like) materials such as blast furnace slag from steel manufacturing or fly-ash from coal-fired power stations. Because pozzolans do not require energy-intensive kiln firing during production their use can reduce environmental and monetary costs.[34]

Fuel consumption for construction processes

While the fossil fuel consumed for transporting building materials is included in embodied energy calculations, the predominant use of construction equipment that run on diesel fuel presents particular issues for the nitrogen cycle. Automotive diesel oil (ADO) has been regarded in some countries as the construction sector's primary delivered energy source. While there is some uncertainty as to the percentage of total delivered energy that ADO comprises, it has been estimated to account for as much as 98% of energy used for construction.[35] While NO_x emissions from petrol engines have been greatly reduced through the introduction of emission abatement technology such as catalytic converters, this technology is not suitable for use in diesel engines. This means that diesel emissions can be highly polluting.

BEEs reduce emissions by using less fuel in general and ADO in particular. On their projects BEEs therefore source materials predominantly from local manufacturers and suppliers to reduce transportation. BEEs also ensure that the construction phase of the project is well scheduled to make most efficient use of plant and equipment. Construction plant and equipment on BEEs' projects are always well maintained and workers are educated to ensure that they understand how to operate machinery to minimise fuel consumption and emissions.

Building and the sulphur cycle

Sulphur is an essential building block for proteins and, in its oxide forms, helps to regulate atmospheric temperature. Sulphur occurs in primeval mineral deposits within the Earth's crust. Sulphur dioxide is released naturally into the atmosphere by volcanic eruption. As a protein builder it enters the environment in various forms, most commonly as an acidic ion (SO_4^-) which is absorbed by plants and

---

Building's major influences on the sulphur cycle

- Fossil fuel consumption
- Building material manufacturing

Major impact

- Acid rain

---

enters the food chain, being released into soils again during organic decomposition.

Phytoplankton in oceans help to release large quantities of SO_2 and SO_4 into the atmosphere through the release of dimethylsulfide (DMS) that oxidises as sea spray evaporates. SO_2 and SO_4 help clouds to form and reflect solar radiation from the upper atmosphere. When sea surface and atmospheric temperatures are warm, phytoplankton produce more DMS, leading to more SO_2 and SO_4 and consequently more cloud and reflection of solar radiation. This reduces the amount of light and heat reaching the Earth's surface, lowering temperatures and reducing DMS production. As the atmosphere and sea surface temperature cool, less DMS is released, less SO_2 and SO_4 form and clouds begin to disappear. Thus less heat radiation is reflected. Sulphur oxides therefore help regulate atmospheric temperatures.

Fossil fuel consumption

The burning of fossil fuels like oil and coal has increased concentrations of atmospheric SO_2 and is the major cause of disruption to this biogeochemical cycle. When oxidised into sulphur trioxide (SO_3) and combined with water particles it forms sulphuric acid (H_2SO_4) and falls as acid rain. Acid rain has been a problem in Northern Europe and the USA where it has caused the degradation of large areas of land and damage to building facades.

Building material production

Sulphur oxides (SO_x) are a key byproduct of manufacturing common building materials like cement, steel and copper. In steel manufacture, for example, sulphur oxides make up a proportion of the 40 kg of emissions including carbon monoxide (CO) and nitrogen oxides (NO_x) produced as byproduct of every tonne of steel. The process of cement manufacture produces between 65 and 240 g of SO_x per tonne of cement. Small amounts of SO_x are also emitted during clay brick manufacture due to release of sulphides in clay.[36] Until recently, the smelting of copper and other suphide ores caused sulphur emissions at levels that devastated the surrounding landscape. The introduction of stringent air quality standards and the availability of emission scrubbing technology has greatly reduced sulphur emissions from these processes.

The disruption of the sulphur cycle is due largely to increasing quantities of atmospheric SO_x as a result of burning oil and coal (in which sulphur is a common impurity) for electricity generation and in engines. Ecologically sustainable building developments therefore help minimise disruptions to the sulphur cycle by minimising life-cycle energy consumption. Maximising efficient use of cement- and metal-based building materials, reducing demand for new materials through recycling and reuse, and minimising transportation requirements are also measures that minimise sulphur cycle disruption.

Building and the water cycle

Water moves from oceans, to atmosphere, through the ecosphere and lithosphere back to oceans in the 'hydrological' cycle. Water evaporates principally from the ocean due to heat from the sun, and is carried through the atmosphere by weather where it accumulates until it condenses and falls as precipitation. Only 10% of all water that falls as rain, falls on land. Water is absorbed by soils and then by plants, used to transport nutrients and is transpired through plant leaves. The processes of evaporation and transpiration are purification mechanisms, trapping many contaminants in soils and cellulose as water is vaporised and returns to the atmosphere. Rainwater collects airborne particulates and gases as it falls, and these are absorbed by plants and soils and transferred to groundwater storage and to streams, rivers, lakes and seas.

Rooftops, pavements and drains

The built environment disrupts the water cycle in many ways. Our urban environment – the houses we live in, the roads we drive on and the building you are in today – has all been imposed on a landscape of estuaries and river deltas, basalt plain, river valley, grass and woodlands. The evolution of these landscapes and ecosystems has created pathways to transfer water from mountains to the sea, and then to the ocean to complete our life-sustaining water cycle.

Building's major influences on the water cycle

Decreased absorption and filtration, and increased flow of run-off due to:

• Vast areas of rooftops, pavement and drains

Major impacts

• Water pollution
• Wasteful consumption

We have replaced this natural cycle with our urban metabolism, the major impacts of which are decreased absorption and filtration, and increased flow of run-off due to vast areas of rooftops, pavement and drains (Fig. 4.4).[37]

Urban areas generally have less porous surface area than a natural environment. This increases surface water run-off. Run-off from streets and rooftops flows into stormwater drains carrying with it litter, oil, soils, fertilisers from gardens, and animal excrement.

Due to the variability of climatic conditions and rainfall patterns, the availability of this water for human use is not guaranteed. In certain areas of Australia, the world's driest populated continent, more than 85% of the rain that falls evaporates before it can become surface run-off available for human use. Areas of Europe and North America on the other hand lose less than 60% of their rainfall to evaporation.[38] Urban areas are generally warmer than their surrounding hinterland due to heat building up in the thermal mass of buildings and paved areas. In most cases water is imported from natural and manmade catchments well beyond the metropolitan area. The water supplied to the city of Los Angeles, for example, travels 400 km from the Owens Valley on California's eastern border,[39] while the entirely urbanised city-state of Singapore imports most of its water via pipelines from Indonesia and Malaysia.[40] Water provided in this way is predominantly used for drinking, washing, and toilet flushing.

Transporting water long distances, and the necessity for siphoning and damming of rivers, obviously disrupts natural water cycles and changes ecosystems where the water is collected. According to the World Resources Institute, people now withdraw more than half of all the water in Earth's rivers.[41] In most cities water used in buildings ends up being discharged into sewers, via treatment plants, and then into rivers, lakes or the sea, adding huge volumes of freshwater, nutrients and persistent chemicals to ecosystems that have not evolved to assimilate these wastes.

The provision of drinking water and sanitation are essential to environmental and social improvement. According to the World Resources Institute, waterborne diseases are the developing world's number one cause of death.[42] However, using the increasingly scarce resource of drinking quality water *for* sanitation is not always an efficient strategy. Unfortunately, this is exactly how the sanitation systems of most of the world's cities work.

Groundwater quality, too, is in decline due to decreased natural filtration and increased volumes of polluted run-off leading to groundwater pollution, and increased scarcity of useable water due to increased harmful chemical and nutrient content.[43,44]

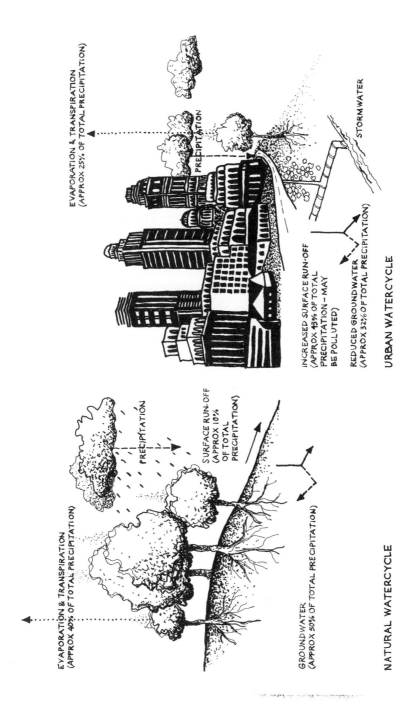

EVAPORATION & TRANSPIRATION
(APPROX 25% OF TOTAL PRECIPITATION)

PRECIPITATION

STORMWATER

INCREASED SURFACE RUN-OFF
(APPROX 43% OF TOTAL
PRECIPITATION – MAY
BE POLLUTED)

REDUCED GROUNDWATER
(APPROX 32% OF TOTAL PRECIPITATION)

URBAN WATERCYCLE

EVAPORATION & TRANSPIRATION
(APPROX 40% OF TOTAL PRECIPITATION)

PRECIPITATION

SURFACE RUN-OFF
(APPROX 10%
OF TOTAL
PRECIPITATION)

GROUNDWATER
(APPROX 50% OF TOTAL PRECIPITATION)

NATURAL WATERCYCLE

Fig. 4.4 The difference between natural and urban water cycles is the plumbing. The natural water cycle on the left retains more groundwater, has lower surface run-off and higher rates of evaporation and transpiration. Urban water cycles increase surface run-off and decrease groundwater because of extensive paving. Run-off is generally not filtered and can pose serious pollution threats.

BEEs take action to minimise water consumption, provide for natural filtration and treatment of water on site, avoid pollution and through recycling, conserve resources.

Building and the phosphorus cycle

Phosphorus is slowly released into the lithosphere through mineral decay. It is absorbed by plants, which are in turn eaten by animals where it is passed through the food chain and is released again into the environment through organic decomposition. A large proportion of phosphorus is leached by groundwater into rivers and to the sea where it dissolves and eventually accumulates in sea-bed sediments. Its role in the functioning of ecosystems is as a catalyst for the absorption and release of energy in cells.

Without phosphorus sunlight could not be converted into energy during photosynthesis, and similarly chemical energy stored in cells could not be released. Animals and plants use the same phosphorus-containing molecule – adenosine triphosphate (ATP) – to drive energy reactions. Plants use ATP as a catalyst in photosynthesis to store energy, while animals use it to release chemical energy stored in cells, a process called cell respiration. ATP is also the primary energy catalyst for immediate physical movement. When you flinch, duck, or sprint you rely on ATP. Olympic sprinters rely almost entirely on their ATP energy to run the 100 m.

Phosphorus is an important catalyst in energy transformations and this is one of the reasons phosphates are such an important fertiliser for agriculture. Soils have varying amounts of natural phosphate. Phosphate-rich soils are very fertile, while low phosphate soils are not. In countries like Australia where the natural soil is very low in phosphate, huge quantities of 'superphosphate' fertiliser are added to soils. The phosphates are imported from mining operations like the one that operated from the 1950s to the 1990s on the island of Nauru, a Pacific island almost entirely made of phosphate. After 40 years of

---

Building's major influences on the phosphorus cycle

- Channelling of phosphate-rich water to the natural environment

Major impacts

- Water pollution

---

mining, 80% of the island has been dug up, shipped off and spread as fertiliser on Australian crop land. Australians and their trading partners have in effect 'eaten' nearly all of Nauru!

Phosphate in waterways

While phosphorus as phosphates may be good in soil they can cause major pollution problems when they build up in water because they stimulate plant growth. Many of the household detergents we regularly mix with water contain phosphates and this is where building enters the phosphorus cycle. As we have discussed previously, the urban water cycle traps and channels far greater volumes of water into natural waterways because paved surfaces and drains do not allow absorption or filtration of surface water.

When rain falls on urban areas phosphates used in garden fertilisers, together with residues from detergents and naturally occurring phosphates, are washed via stormwater drains into waterways. Under the right conditions phosphates can cause algal blooms and a build-up of plant life, which can reduce stream flows, increase siltation and decrease ecological health. A similar situation can occur where sewage effluent is discharged into aquatic environments because of the use of phosphate-containing detergents for washing. Our urban systems therefore help speed up the transmission of phosphate from source to sink.

BEEs reduce impacts on the phosphorus cycle by reducing the volume of stormwater being discharged into stormwater drains. They also reduce areas of non-porous paving to provide natural filtration and, wherever possible, landscape using plant species needing no added fertiliser to thrive in the soil found on-site.

BEEs also minimise water consumption. The lower the amount of water used for sanitary purposes, the lower the risk of environmental consequences should detergents with phosphates be used. Recycling grey water (water that has been used for general washing) for irrigation is another ecology friendly building strategy (see Fig. 4.5).

Large-scale agriculture has caused a far greater direct impact on the functioning of the phosphorus cycle than urban environments, due to the extensive use of superphosphate fertilisers. As urban areas are by far the greatest importers of produce grown in this way, areas for local community food production within building developments are another strategy that BEEs consider. Local food production also reduces demand for, among other things, transport fuel, refrigeration and packaging and therefore has a positive influence on a number of other biogeochemical cycles.

Fig. 4.5 A rainwater collection system in a suburban backyard becomes a water feature as well as an irrigation channel. Photograph by Andrew Wood.

Impacts on ecosystems

It may be hard to visualise an office tower or apartment block in the middle of a big city, houses in a sprawling suburb or an informal settlement as a part of an ecosystem. But have you ever stopped to consider where water is flowing from, where the materials were mined or where the garbage and sewage is going? Have you then wondered whether the ecosystems, of which our buildings are a part, and from which our buildings and the resources that feed them come from, and into which our waste streams flow, are being damaged?

Ecosystems are characterised as systems of biological communities in their physical environments. Ecology, the study of ecosystems, has discovered that they consist of vast and complex systems of relationships between species and between species and their non-living environments. The complex systems of relationships in ecosystems, referred to by Fritjof Capra[45] among others as 'The Web of Life', have evolved to transfer energy, matter and information so that the functioning of the whole system can be maintained.

Ecosystems in this way provide all of the goods and services that we depend on to survive.[46] They therefore provide models for us to emulate as we design and build sustainable human systems. In order to understand how ecosystems work, ecologists concentrate their study on the functioning of *communities* of species and the *systems* of *relationships* between communities, rather than on the individual habits and behaviour of animals or organisms. Adopting this *systems perspective*, rather than focusing on the functioning of individuals, helps ecologists to understand how the whole system works.[47]

Ecosystems are of course already infused with human systems. The farmland upon which we grow our food, tree plantations, public parks and gardens, even our own backyards, are elements of ecosystems. These types of 'managed' ecosystems have been modified in order to enhance the yield of particular goods or services, while 'natural' ecosystems like forests, oceans and range lands retain much of their original complexity and function. Yet because of our interdependency, both 'natural' and 'managed' ecosystems are affected by human activity.

We rely on so many different ecosystems to provide the goods and services that support both our lives and our lifestyles. BEEs concern themselves to ensure that the effects of our interaction with ecosystems support rather than undermine their life-supporting functions. BEEs make buildings that have minimal negative ecological impacts by understanding how buildings are elements of ecological systems.

Characteristics of ecosystems

In ecology there have been two competing views on the characteristics of ecosystems. One view assumes nature is constant and that ecosystems always develop toward a diverse stable equilibrium state. The other holds that while ecosystems develop diversity, they can do this along many pathways and therefore move between multiple equilibrium states.[48] In recent times the steady state perspective has been shown to be incomplete[49] and the view of ecosystems as dynamic is more widely accepted. According to this view ecosystems are said to have a number of important qualities. Typical qualities of ecosystems include diversity and resilience, positive and negative feedback, emergence, self-organisation, unpredictability and hierarchical organisation. Ecosystems exhibit a wide range of variation in complexity and rates of change. Understanding the impacts we have on ecosystems is therefore a complex matter. However, with a basic introduction to how ecosystems operate, we can identify some important issues.

Diversity and resilience

Ecosystems are thought to develop diversity as they mature. Diversity is a key mechanism of *ecosystems resilience*[50] or the ability of an ecosystem to maintain its functions while coping with change. Ecological diversity relates to the species of plants and animals that exist in an ecosystem and to all elements within the ecological hierarchy. Diversity exists at genetic, species, habitat and niche levels within ecosystems.[51] Each type of diversity is important in contributing to the system's ability to cope with change.

- *Genetic diversity* refers to the total number of genes in an ecosystem. Species can develop different genetic strains, which increase the adaptability of the species to different environmental conditions. Lack of genetic diversity can be potentially ecologically and economically devastating. Modern agriculture, for example, has reduced the genetic diversity of some domestic crops and animals due to globalised market demand for homogenous products. This has led to increased vulnerability of crops to disease and insect attack.[52]
- *Species diversity* or species richness refers to the number of different species in an area. Some ecosystems are species-rich and some are less so, however the biological importance of an area is influenced by the endemicity (uniqueness) of the organisms that live there, as well as by the total number of species. Tropical forests are an example of species-rich ecosystems, containing at least 50% of all known species on Earth. Less species-rich ecosystems include grasslands and desert areas, although the species that live in these areas may be highly specialised and unique.
- *Habitat diversity* describes the number of different physical environments or 'micro-climates' that exist in an ecosystem. Habitats will be populated by organisms that may have adapted specifically to that location. The biodiversity of an ecosystem is largely a result of its habitat diversity. Human induced species extinction is largely due to habitat destruction, degradation and fragmentation.
- *Niche diversity* refers to the number of specialised relationships that occur between organisms and their habitat. Niche diversity

is related to habitat and species diversity and is threatened most directly by habitat loss.

Ecosystems develop *resilience* or the capacity to adapt non-destructively to change through increasing diversity of ecosystem elements such as genes, species and populations of species.[53] Research has shown that ecosystems that have greater diversity have a greater capacity to absorb change. This is due to the complex interrelationship between species operating at different scales of influence within the system.[54] The resilience of an ecosystem relates to its carrying capacity and therefore sets parameters of buffer zones within which impacts can be absorbed. Accumulation of small impacts over time, or the imposition of large impacts that are within the system's carrying capacity, may not change its state, but once at its limit, even small additional impacts can breach these buffer zones and cause a rapid and catastrophic 'flip' of an ecosystem to a different, normally less diverse, state.[55] Changes of state are common in ecosystems. The dynamic emergent properties of ecosystems have been described as self-organisation. This behaviour is partially explained by the influence of feedback.

Feedback, emergence and self-organisation

Relationships between living and non-living components of ecosystems are such that they provide mechanisms for recycling or 'feedback' of matter and energy. Ecosystems survive by constantly cycling matter from one organism to the next, and then to storage in non-living components such as soils, water or the atmosphere. This cycling requires the constant input of solar energy. Accurately predicting the effects of human induced changes in an ecosystem is therefore impossible using a linear cause and effect approach. (See Chapter 5 for more discussion.) Instead, a systems approach that considers non-linear hierarchical relationships and ecological carrying capacities needs to be applied.[56]

Ecosystems evolve and survive as a consequence of constant feedback of energy and matter between organisms and their environment. This should not be thought of as a static feedback loop because the energy and matter fed back to the organism are not exactly the same as the energy and matter it originally used. The organism therefore adapts to the new inputs. Thus the conditions of life constantly change. As Magulis states in her explanation of the *Gaia hypothesis*:

'Life actually makes and forms and changes the environment to which it "adapts". Then that "environment" feeds back on the life that is changing and acting and growing in it. There are continuous cyclical interactions.' [57]

The Gaia hypothesis applies to global systems, and ecosystems, despite representing fractions of the global system, are thought to display the same dynamic and self-regulating propensities. As Holling[58] puts it 'Ecosystems hence, are GAIA small writ'.

Hierarchical

Ecosystems can be categorised hierarchically according to their biodiversity.[59] Understanding which category of ecosystem is likely to be affected by a building decision provides a way of assessing its likely environmental implications. The categories as suggested by Lamb, arranged from most to least diversity, are:

(a) Mature (natural) ecosystems: such as old-growth forest, wetlands, grassland and established coral reefs;
(b) Immature (natural) ecosystems: areas recovering from damage or regenerating;
(c) Simplified (natural) ecosystems: such as selectively logged forest, grazed grassland, and areas under controlled burning;
(d) Mixed artificial ecosystems: such as land under crop rotation, parks, gardens, agro-forestry;
(e) Monoculture (artificial) ecosystems: such as timber plantations and agricultural crops;
(f) Zero culture (artificial) ecosystems: such as urban areas, and open-cut mines.

Most building draws resources from each of these types of ecosystem, and generally has the effect of reducing biological diversity in all of them. In other words building activity moves ecosystems from level (a) in the hierarchy towards level (f). This reduction in diversity can also reduce the resilience of an ecosystem. Reducing diversity reduces the carrying capacity of ecosystems and increases their instability.[60] Protecting biodiversity is therefore fundamental to sustaining ecosystems and for maintaining the ecological carrying capacity required to accommodate human activity. BEEs takes steps to increase biological diversity from levels (d), (e) and (f) towards level (a).

Biodiversity

The diversity of a natural ecosystem can change over time from sim-plified and immature states containing little diversity to fully mature states containing sophisticated diversities. Ecosystems can also be human constructed, ranging from zero culture and monoculture ecosystems to mixed artificial ecosystems. Biodiversity is important for life on Earth because it:

* represents accumulated genetic history and evolution;
* contributes to the resilience of ecosystems;
* supports cultural and technical advances derived from nature; and
* contributes to human inspiration and peace.

Genetic history

The diversity of life contains Earth's genetic history and therefore the building blocks of evolution. The story of evolution on this planet is not just one of learning from 'mistakes' through mutation (muta-tion is the least successful evolutionary tool at nature's disposal). The evolution of life has progressed more through the combination, networking and collaboration of many forms of organism over time. The story of evolution then is also one of 'continual cooperation and mutual dependence of all life-forms'[61] rather than simply a matter of 'survival of the fittest'.

Bacteria, for example, directly trade genes in a process called DNA recombination (a process we are now learning to mimic through biotechnology) in addition to mutation. For complex multicellular creatures like humans, mutation has proven an unreliable evolution-ary tool and we are unable to swap DNA like bacteria. Our major evolutionary tool is the process of symbiosis –the tendency to live in close proximity to, or share our bodies with, other organisms.[62]

Maintaining biodiverse ecosystems provides the opportunity for us to continue to find out more about how evolution works, and also maintains a pool of genetic resources that assures its continuity. Eco-systems may not begin restructuring processes after they have been damaged unless there is sufficient biodiversity with a memory of how to rebuild.[63]

Mechanism of resilience

The richness and complexity of a biological community is the key to its resilience. Increased diversity at each scale within an ecosystem increases its ability to continue to function despite changes in environmental conditions. An ecosystem, as a network structure, has many elements with overlapping functions. Should one species be destroyed by a severe disruption, another species can partially replace it. Thus a diverse ecosystem can reorganise itself and continue to function.[64]

Source for technical advances

Humans have derived many direct benefits from the biodiversity of Earth's ecosystems. All of our food and many of our fibres and medicines come from other organisms. However, despite the incredibly large number of organisms upon which humans could draw for these purposes, we currently use a very limited variety. Only 20 of the approximately 200 domesticated varieties of food crops, for example, are major sources of food, with two families – Gramineae (grasses, including cereals) and Leguminosae (legumes, including peas, beans and lentils) – predominating.[65]

Scientists have discovered only 1.7 million of an estimated 12.5 million species living on Earth. Loss of biodiversity results in lost opportunities for discovering new technologies of human and perhaps global significance. With growing populations, decreasing food and water and increasing environmental pollution, maintaining biological diversity means maintaining possibilities for adapting to the global changes that lie ahead.

Ecosystems provide essential life-support goods and services such as clean air, fresh water, food and waste assimilation. In biological terms these services are essential for survival. As of the year 2000, activities that directly rely on ecological health such as agriculture, fishing and forestry were providing 50% of jobs worldwide, and contributing more to the economies of 25% of all nations.[66]

Our industries and our economic well-being therefore rely on these ecological services. Recent research by the University of Maryland in the US calculated the value of ecosystems services as an element of the total economic value of the planet. The results are summarised in Table 4.2.[67] According to the study, the value of ecological services to the global economy is about $US42 trillion, more than twice the value of total global GDP.

Table 4.2 'Economic value' of ecosystems.[67]

Ecosystem	Worth (US$ pa)	Function
Forest	6 trillion	• regulates greenhouse gases and water flows • prevents erosion • forms soil • detoxifies the environment • provides timber and medicines • recreation and cultural value
Grass/rangelands	1.16 trillion	• stops soil erosion • provides pollinators so plants can reproduce • forms soil
Wetlands	6.24 trillion	• store and retain water • control floods • protects the coast from storms
Coral reefs	475 billion	• controls pollution • treats waste • protects coasts from storms • provides for tourism
Lakes/rivers	2.2 trillion	• store and retain water • control floods • provides fish for food
Open ocean	10.74 trillion	• generates oxygen and absorbs carbon dioxide • provides fish for food
Other	15.2 trillion	

Source: *Nature Magazine.* In G. Bryar (1997) Ecology put in economic terms. *Herald Sun* 19 May 1997, p.21.

Contributor to inspiration and peace

The complexity and diversity of nature has had a profound influence on the human mind and soul. Indigenous cultures around the world have, over thousands of years of direct contact with and in nature, developed sophisticated stories, myths, lore and technology embedded with profound ecological wisdom. Western science too, has been inspired by the complexities and interdependencies of biodiversity:

'Science's bright, "objective" light has provided us with increasingly elegant conceptual mirrors of nature. It has teased away a number of the vital living threads that bind the biosphere's intricate, interconnected flows of energy and matter into a single ecological whole. It has revealed this planet as the shared birth-place of all species and our only true home.'[68]

All elements of Earth, whether living or non-living, are interdependent. In this context biodiversity perhaps represents the richness of the experience of life because, when immersed in a biodiverse landscape, we are surrounded by many reminders of our part in the whole, of how we are connected.

To understand what to do to protect biodiversity we need to know what threats building activity pose.

Building and biodiversity

'If people suddenly disappeared from the Earth, the planet could recover and within 1,000 years, it would look as it did 100,000 years ago. If insects disappeared nothing on land would survive.' Rafael Gonzalez, Mexico – UNEP[69]

We fundamentally rely for survival upon the incredible diversity of living things on Earth. Loss of biological diversity is considered to be one of the major environmental threats to continued human survival and opportunity.

'The health and survival of Earth and its people depends on the operation of the natural systems which sustain the earth's processes and which in turn depend on the ability to change. The most important parts of these intercon-nected systems are the biological parts.'[70]

The International Union for the Conservation of Nature (IUCN) reported that 25% of the world's mammals (1158 species) and 11% of birds (1065 species) are threatened with extinction.[71] The statistics are worse at a national level with countries like the US reporting vast numbers of species as threatened with extinction. Figures from the US indicate that almost 70% of mussels, 50% of crayfish and 37% of freshwater fish are threatened due to pollution and habitat loss.[72] Australia's rate of mammal extinction is the world's worst, having lost 10 of 144 species of marsupials and 8 of 53 species of rodents in the last 200 years. In Australia, 5% of higher plants, 9% of birds, 16%

of amphibians and 23% of mammals are either extinct, endangered or vulnerable.[73]

The building industry and the increasing urbanisation of human settlements pose major threats to the diversity of life through contributing both directly and indirectly to habitat loss, degradation and fragmentation. Ecologically sustainable building projects take steps to protect biodiversity and, where possible, repair damaged ecosystems and increase the diversity of life. To understand why it is important to take these steps, and what kinds of approaches are available to building professionals, a basic knowledge of concepts and implications of biological diversity is required.

Biological diversity is a necessary aspect of resilient ecosystems and provides humans with the biological resources required to support our ways of life and our building industries. It has been estimated that twelve domesticated and twelve wild varieties of plant species are commonly used directly or as an ingredient in building materials and products in the Northern Hemisphere. The building industry in Europe relies predominantly on two varieties of conifer and eleven varieties of deciduous trees including oak, beech, ash and poplar.[74] The Australian building industry draws on thirty-four different types of eucalypt and eleven varieties of conifer[75] Animals too have played their part, and indeed have had their parts used in the manufacture of building materials, particularly in vernacular buildings. Animal products like skins, hair, wool, hooves and fat have been widely used for everything from wall cladding to paint and adhesives.

The building industry relies on the variety of living things on Earth for important building materials like timber and natural fibres, and for providing essential services that clean up wastes like sewage, stormwater and organic matter, purify air and soil and maintain healthy drinking water. Yet biological diversity on Earth is in decline. So what role is building playing in loss of biodiversity?

Biodiversity is threatened mainly by loss, degradation, and fragmentation of habitat, species extinction, introduction of exotic species, and genetic assimilation. Of these threats the building industry contributes most to negative changes in habitat through extraction of resources – particularly timber – and changing land-use, from rural to urban, as well as water extraction and pollution.

Extraction of resources for building

The construction industry is one of the largest consumers of resources, requiring major inputs of energy, materials and land. Biodiversity of areas in which minerals for buildings are derived are initially completely destroyed by mining operations and potentially by mine wastes and tailings (see previous section on ecological footprints and rucksacks). The extraction of most metals and minerals used in building has severe consequences on the diversity of ecosystems. Some materials, particularly heavy metals like zinc, copper, chromium, mercury, cadmium and lead, persist in the environment after they have been extracted and processed and can build up in ecosystems over time.

Timber is a major building material whose use has potentially devastating consequences on biological diversity, particularly when it is sourced from old growth and tropical forests (see Fig. 4.6). Forest ecosystems are home to about two-thirds of all known species, have the greatest species diversity of any type of ecosystem, and contain the largest number of threatened species.

The building industry uses approximately 875 million cubic metres of timber in construction globally, which accounts for 25% of global wood consumption annually.[76] In the USA, Scandinavia and Australia timber is the predominant building material for domestic construction.[77-79]

Knowing the source of this massive consumption of timber is key to determining the impact of timber on biodiversity. As a rule of thumb, if the timber is a species that comes from a rainforest or an old-growth forest then it will have a far greater impact on the Earth's remaining natural biodiversity than timber from a plantation. Countries like the US[80] and the UK[81] import the majority of the timber they require. It is therefore harder to ensure that timber being used is coming from a sustainably managed source. Organisations such as the Forest Stewardship Council[82] provide certification for sustainable timbers.

Using plantation timber can help reduce the impact of logging in native old-growth and tropical forests and therefore can help protect indigenous habitats and biodiversity. While the use of plantation timber has its environmental benefits, the plantation industry has some environmental issues of its own that need to be considered. The benefits and drawbacks of plantation timber are described in Table 4.3.[83,84]

BEEs always try to select timber that is produced in an ecologically sustainable way. A rule of thumb they follow is to always avoid specifying or purchasing rainforest or old-growth timber. To ensure that timber they are using is being produced sustainably they can refer to certification organisations such as the Forest Stewardship Council.

Fig. 4.6 Old-growth biodiversity and clearcut devastation.

Table 4.3 Benefits and drawbacks of plantation timber.[83,84]

Plus: Plantation timber is a renewable resource	Natural forests are a forest resource. They provide many services besides the timber that grows in them. If they are cut down, the forest resource is gone. Plantations on the other hand are basically tree farms. Their service is the timber in the trees that grow there.
Plus: Harvesting plantation timber poses little threat to biodiversity	Timber plantations provide little habitat for indigenous plants and animals. Harvesting timber from plantations therefore does not destroy diverse habitat or species rich areas. The establishment of plantations on the other hand may be devastating to ecosystems in areas where they replace old growth forest or other diverse ecosystems.
Minus: Plantation timber is monoculture	Plantations are usually made up of one species of tree, have very little undergrowth and they are cut relatively young before trees have developed hollows for animals to nest in. Even in areas where plantation species are native, plantations lack the diversity of habitat required for many indigenous animals.
Minus: Plantation timber is relatively energy intensive	Timber plantations require more energy intensive infrastructure than native forests. Plantation trees are propagated in nurseries prior to their introduction into the plantation. Fertilisers, herbicides and pesticides are also used during the establishment of seedlings to encourage growth, control competing plants and to kill predators. The quantities of these additives are, small when compared with traditional agriculture, however, risks of environmental pollution are posed by potential leaching of chemicals through ground water as well as loss of contaminated topsoil due to erosion after logging.
On the whole	The benefits of using plantation-grown timber to conserve biodiversity far outweigh the disadvantages. The overriding rules to protect biodiversity when using timber are to recognise its ecological value and use it as efficiently and effectively as possible.

Source: Based on Grey, A. & Hall, A. (1999) *Forest Friendly Building Timbers* Earth Garden, Trentham, Australia, and Lawson, B. (1996) *Building Materials, Energy and the Environment – Towards Ecologically Sustainable Development* Royal Australian Institute of Architects, Red Hill, Australia.

Biodiversity loss through changing land use

'Singapore in its founding year of 1819 was a tropical island covered in rain forest, encircled by sandy beaches and sheltered mangrove shorelines. This initial state of rich biodiversity was soon shattered by man's activities … In more recent times most rivers were dammed for reservoirs or concrete culverted and long lengths of shoreline were reclaimed. Construction activity resulted in many hills being flattened quarries excavated and primary forests and mangroves were lost. Today over 50% of the main island is urbanised and it is anticipated that by 2010 this will be 75%.'[85,*]

As this description of Singapore's establishment shows, when land is converted from its natural state to rural or urban uses, there is generally a massive decline in biodiversity in the area directly affected by the change in use. Increasing impacts on remote ecosystems also occur as a result of the patterns of resource consumption evident, particularly in urban areas (see discussion on ecological footprints).

Urban areas have over time replaced functioning ecosystems with very simplified ecosystems that are made up of small pockets of remnant indigenous plants and animals and, often larger populations of introduced species. In Brisbane Australia for example, only 600 hectares of 6000 hectares of pre-settlement rain forest has survived since the city's establishment in the 1840s. In the Sydney region less than one percent of the original blue gum high forests, and less than 0.5 percent of turpentine-ironbark forests remain.[86] Of particular concern in Australia is habitat loss in coastal areas as a result of major urban encroachment.

With 60% of humans living within 100 kilometres of the coastline, it is not surprising that approximately 51% of the world's coastal ecosystems for example, are at 'significant risk of degradation' according to the World Resources Institute.[87] The World Resources Institute reported in 1998 that Europe's coastal ecosystems are under greatest pressure, followed by Asian coastal ecosystems.[88] In many countries the pressures on coastal ecosystems are being added to directly by building activity. In the UK for example, reports suggest an increasing proportion of the 300 million tonnes of aggregate consumed each year is being quarried in coastal areas,[89] while in Asian Countries like Hong Kong and Singapore coastal areas are commonly 'reclaimed' for use as building sites. Inland cities place similar pressures on river ecosystems, and have in some cases depleted forests and sources of drinking water.[90]

*Figures quoted by Briffet *et al.* (reference 85) do not include green space or nature reserves, they only refer to land that is actually built upon. Singapore is regarded as 100% urbanised by both the World Resources Institute and the UN Habitat program.

References

1 Woodwell, G.M. (1971) Effects on pollution on the structure and physiology of ecosystems. In: Matthew, W.H., Smith, F.E. & Goldberg, E.D. (eds) *Man's Impact on Terrestrial and Oceanic Ecosystems*. MIT Press, Cambridge, MA, USA. pp.47–58.

2 Ramachandran, A. (1990) The impact of construction technology on the environment. Keynote address XVIII IAHS World Congress, October, Rio de Janeiro.

3 Australian Bureau of Agricultural and Resource Economics (ABARE) (1994) *Quarterly forest products statistics – December quarter 1994*. Australian Government Publishing Service, Canberra.

4 Yencken, D. & Wilkinson, D. (2000) *Resetting the Compass: Australia's journey towards sustainability*. CSIRO Publishing, Melbourne. pp.125–129.

5 Clough, R. (1994) Environmental Impacts of Building Construction. *Buildings and the Environment* – Proceedings of the first International conference of CIB task group eight. Building Research Establishment, Watford, UK. 16–20 May.

6 Lawson, B. (1996) *Building Materials, Energy and the Environment: towards ecologically sustainable development*. Royal Australian Institute of Architects, Red Hill Australia.

7 Meadows, D., Meadows, D., Randers, J. & Behrens, W. (1972) *The Limits to Growth: A Report for the Club of Rome's Project on the Predicament of Mankind* Universe Books New York.

8 Yencken, D. & Wilkinson, D. (2000) *Resetting the Compass: Australia's journey towards sustainability*. CSIRO Publishing, Melbourne.

9 World Watch Institute (2000) Earth Day 2000 – A 30 year report card. *World Watch* March/April. World Watch Institute Washington DC, USA. pp.10–11.

10 Fuller, R.B. (1981) *Critical Path*. St. Martin's Press, New York.

11 Howard, N. (2000) *Sustainable Construction: the data*. CR 258/99. Building Research Establishment Centre for Sustainable Construction, Watford, UK. p.14.

12 Cunningham, P. & Saigo, B. (1997) *Environmental Science: a global concern*. Fourth Edition. McGraw Hill, New York. p.56.

13 UNEP (2001) Summary for Policy Makers – A report of Working Group One of the Intergovernmental Panel on Climate Change. *Climate Change Synthesis Report* United Nations, Geneva, Switzerland.

14 UNEP (2001) Summary for Policy Makers – A report of Working Group One of the Intergovernmental Panel on Climate Change. *Climate Change Synthesis Report* United Nations, Geneva, Switzerland.

15 Lawson, B. (1996) *Building Materials, Energy and the Environment: towards ecologically sustainable development*. Royal Australian Institute of Architects, Red Hill, Australia. p.50.

16 Van Oss, H. (1999) Cement. *US Geological Survey Minerals Yearbook – 1999* Vol 1. Minerals and Metals. US Government Printing Office, Washington, USA. pp.16.1-16.13. http://minerals.usgs.gov/minerals/pubs/commodity/myb/. Accessed 4/02/02.

17 Van Oss, H. (1999) Cement. *US Geological Survey Minerals Yearbook – 1999* Vol 1. Minerals and Metals. US Government Printing Office, Washington, USA. pp.16.1-16.13. http://minerals.usgs.gov/minerals/pubs/commodity/myb/. Accessed 4/02/02.

18 Lawson, B. (1996) *Building Materials, Energy and the Environment: towards ecologically sustainable development.* Royal Australian Institute of Architects, Red Hill Australia.

19 Schmidt-Bleek, F. (1994) *Carnoules declaration of the Factor Ten Club.* Wuppertal Institute. Cited in: Weizsäcker, E. von, Lovins, A.B. & Lovins, L.H. (1997) *Factor four: doubling wealth - halving resource use: the new report of the Club of Rome.* Allen & Unwin, Sydney, Australia.

20 Van Oss, H. (1999) Cement. *US Geological Survey Minerals Yearbook – 1999* Vol 1. Minerals and Metals. US Government Printing Office, Washington, USA. pp.16.1-16.13. http://minerals.usgs.gov/minerals/pubs/commodity/myb/. Accessed 4/02/02.

21 Van Oss, H. (1999) Cement. *US Geological Survey Minerals Yearbook – 1999* Vol 1. Minerals and Metals. US Government Printing Office, Washington, USA. pp.16.1-16.13. http://minerals.usgs.gov/minerals/pubs/commodity/myb/. Accessed 4/02/02.

22 Van Oss, H. (2001) *Cement* US Geological Survey Mineral Commodity Summaries, January. Government Printing Office, Washington, U.S.A. pp40. http://minerals.usgs.gov/minerals/pubs/commodity/myb/. Accessed 4/02/02.

23 Marland, G., Boden, T.A. & Andres, R.J. (2001) Global, Regional, and National Fossil Fuel CO_2 Emissions. In: *Trends: A Compendium of Data on Global Change.* Carbon Dioxide Information Analysis Center, Oak Ridge National Laboratory, US Department of Energy, Oak Ridge, Tenn., USA.

24 Ecologically Sustainable Development Working Groups (1991) *Final Report – Energy Use*, AGPS, Canberra, Australia. Cited in: Lawson, B. (1996) p.12.

25 Roodman, D. & Lenssen, N. (1995) *A Building Revolution: how ecology and health concerns are transforming construction.* World Watch Paper No. 124, March. World Watch Institute, Washington DC, USA. p.23.

26 Friends of the Earth Europe (1995) *Towards Sustainable Europe.* Friends of the Earth Europe, Brussels.

27 Commonwealth of Australia (AGO) (1999a) *Australian commercial building sector greenhouse gas emissions 1990–1910, executive summary report.* Australian Greenhouse Office, Canberra, Australia. p.13.

28 Commonwealth of Australia(1999b) *Australian residential building sector greenhouse gas emissions 1990–2010 final report.* Australian Greenhouse Office, Canberra, Australia.

29 Kramer, D. (2000) Nitrogen. *US Geological Survey Minerals Yearbook – 2000* Vol 1. Minerals and Metals. US Government Printing Office, Washington, USA. pp.55.1–16.12. http://minerals.usgs.gov/minerals/pubs/commodity/myb/. Accessed 4/02/02.

30 Cunningham, P. & Saigo, B. (1997) *Environmental Science: a global concern.* Fourth Edition. McGraw Hill, New York.

31 World Resources Institute (2002) *Earth Trends the Environmental Information Portal* – Climate & Atmosphere Data Tables http://earthtrends.wri.org/searchable_db/index.cfm. Accessed 04/02/02.

32 Canadian Centre for Mineral and Energy Technology & Radian Canada Incorporated (1993) *Raw Materials Balances, Energy Profiles and Environmental Unit actor Estimates for Cement and Structural Concrete Products*, Forintek Canada Corporation. p.47.

33 Curwell, S. & Cooper, I. (1998) The implications of urban sustainability. *Building Research and Information* **26** (1), 17–28.

34 Van Oss, H. (2001) *Cement.* US Geological Survey Mineral Commodity Summaries, January. Government Printing Office, Washington, USA. http://minerals.usgs.gov/minerals/pubs/commodity/myb/. Accessed 4/02/02.

35 Bush, S., Holmes, L. & Luan, H.T. (1995) *Australian energy consumption and production: historical trends and projections to 2009–10.* Australian Bureau of Agricultural and Resource Economics Research Report No. 95.1. Canberra, Australia.

36 Lawson, B. (1996) *Building Materials, Energy and the Environment: towards ecologically sustainable development.* Royal Australian Institute of Architects, Red Hill, Australia.

37 Melbourne Water (2001) Melbourne Water's Waterways and Drainage facts and figures. *Infostream.* Melbourne Water, January. Melbourne, Australia.

38 Crabb, P. (1997) *Impacts of anthropogenic activities, water use and consumption on water resources and flooding, Australia:* State of the environment technical paper series (Inland Waters). Environment Australia, Department of the Environment, Canberra. Australia.

39 Cunningham, P. & Saigo, B. (1997) *Environmental Science: a global concern.* Fourth Edition. McGraw Hill, New York. p.426.

40 US Department of Commerce (2000) *National Trade Data Bank*, Published November 3, 2000 at http://www.tradeport.org/ts/countries/singapore/isa/isar0022.html. Accessed 11/02/02.

41 United Nations Development Program (UNDP), United Nations Environment Program (UNEP), World Bank & World Resources Institute (2000) *A Guide to World Resources 2000–2001: people and ecosystems: the fraying web of life.* World Resources Institute, Washington DC.

42 World Resources Institute (1998) *World Resources Report 1998–1999 Environmental Change and Human Health.* Oxford University Press, Oxford, UK.

43 United Nations Development Program (UNDP), United Nations Environment Program (UNEP), World Bank & World Resources Institute (2000) *A Guide to World Resources 2000–2001: people and ecosystems: the fraying web of life.* World Resources Institute, Washington D.C.

44 Sampat, P. (2000) Groundwater shock: the polluting of the world's major freshwater stores. *World Watch.* World Watch Institute Washington DC, USA. **13** (1) 10–22.

45 Capra, F. (1997) *The Web of Life: a new synthesis of mind and matter.* Flamingo, Harper Collins, London.

46 United Nations Development Program (UNDP), United Nations Environment Program (UNEP), World Bank & World Resources Institute (2000) *A Guide to World Resources 2000–2001: people and ecosystems: the fraying web of life.* World Resources Institute, Washington DC.

47 Cunningham, P. & Saigo, B. (1997) *Environmental Science: a global concern.* Fourth Edition. McGraw Hill, New York. p.52.

48 Holling, C. (1986) The resilience of terrestrial ecosystem: local surprise and global change. In: Clark, W. & Munn, R. (eds) *Sustainable Development of the Biosphere.* Cambridge University Press, Cambridge, UK.

49 Lister, N. & Kay, J. (1999) Celebrating Diversity: adaptive planning and biodiversity conservation. In: Bocking S. (ed.) *Biodiversity in Canada: an introduction to environmental studies.* Broadview Press, Peterborough, Canada. pp.189–218.

50 Lister, N. & Kay, J. (1999) Celebrating Diversity: adaptive planning and biodiversity conservation. In: Bocking S. (ed.) *Biodiversity in Canada: an introduction to environmental studies.* Broadview Press, Peterborough, Canada. pp.189–218.

51 Lamb, R. (1995) Biodiversity. In: *BDP Environment Design Guide.* General Issues Paper 3, May. Melbourne, Australia.

52 United Nations Environment Program (UNEP), and Peace Child International (1999) *Pachamama, our earth – our future, by young people of the world.* Evans Brothers Ltd, London.

53 Lister, N. & Kay, J. (1999) Celebrating Diversity: adaptive planning and biodiversity conservation. In: Bocking S. (ed.) *Biodiversity in Canada: an introduction to environmental studies.* Broadview Press, Peterborough, Canada. pp.189–218.

54 Peterson, G., Allen, C. & Holling, C. (1998) Ecological Resilience, Biodiversity, and Scale. *Ecosystems* Vol 1. New York, Inc. USA. pp.6–18.

55 Kay, J. (2002) On complexity theory, exergy and industrial ecology. In: Kibert, J., Sendzimir, J. & Guy, B. (2002) *Construction Ecology: Nature as the basis for green buildings.* Spon Press, New York. pp.72–107.

56 Kay, J. (2002) On complexity theory, exergy and industrial ecology. In: Kibert, J., Sendzimir, J. & Guy, B. (2002) *Construction Ecology: Nature as the basis for green buildings.* Spon Press, New York. pp.72–107.

57 Margulis, L. (1989) 'Gaia: The Living Earth' – Interview with Fritjof Capra, in *The Elmwood Newsletter* Berkeley, Cal., Vol. 5, No.2. Cited in: Capra, F. (1997) *The web of life – A new synthesis of mind and matter.* Flamingo, Harper Collins, London. p.106.

58 Holling, C. (1986) The Resilience of Terrestrial Ecosystem: local surprise and global change. In: Clark, W. & Munn, R. (eds) *Sustainable Development of the Biosphere.* Cambridge University Press, Cambridge, UK. p.298.

59 Lamb, R. (1995) Biodiversity. In: *BDP Environment Design Guide* General Issues Paper 3, May. Melbourne, Australia.

60 Peterson, G., Allen, C. & Holling, C. (1998) Ecological Resilience, Biodiversity, and Scale. *Ecosystems* Vol 1. New York, Inc. USA. pp.6–18.

61 Capra, F. (1997) *The Web of Life: a new synthesis of mind and matter.* Flamingo, Harper Collins, London. p.226.

62 Margulis, L. & Sagan, D. (1995) *What is Life?* Simon & Schuster, New York.

63 Lister, N. & Kay, J. (1999) Celebrating Diversity: adaptive planning and biodiversity conservation. In: Bocking S. (ed.) *Biodiversity in Canada: an introduction to environmental studies.* Broadview Press, Peterborough, Canada. pp.189–218.

64 Peterson, G., Allen, C. & Holling, C. (1998) Ecological Resilience, Biodiversity, and Scale. *Ecosystems* Vol 1. New York, Inc. USA. pp.6–18.

65 United Nations Environment Program (UNEP), and Peace Child International (1999) Op.Cit.

66 United Nations Development Program (UNDP), United Nations Environment Program (UNEP), World Bank & World Resources Institute (2000) *A guide to world resources 2000–2001: people and ecosystems: the fraying web of life.* World Resources Institute, Washington D.C.

67 *Nature Magazine.* In G Bryar (1997) Ecology put in economic terms. *Herald Sun* 19 May 1997, p.21.

68 Knudtson, P. & SuzU.K.i, D. (1992) *Wisdom of the elders.* Allen & Unwin, Sydney.

69 United Nations Environment Program (UNEP), & Peace Child International (1999) *Pachamama, our earth – our future, by young people of the world.* Evans Brothers Ltd, London.

70 Lamb, R. (1995) Biodiversity. In: *BDP Environment Design Guide* General Issues Paper 3, May. Melbourne, Australia.

71 United Nations Environment Program (1999) *Global Environment Outlook – GEO 2000* United Nations Environment Program, Nairobi, Kenya.

72 Master *et al.* (1998) – United Nations Environment Program (1999) *Global Environment Outlook – GEO 2000* United Nations Environment Program, Nairobi, Kenya.

73 Alexander, N. & Taylor, R. (eds) (1996) *State of the Environment Australia: Executive Summary.* CSIRO Publishing, Collingwood, Victoria, Australia.

74 Alexander, N. & Taylor, R. (eds) (1996) *State of the Environment Australia: Executive Summary.* CSIRO Publishing, Collingwood, Victoria, Australia. pp.164–165.

75 Willis, A.M. & Tonkin, C. (1998) *Timber in Context: a guide to sustainable use.* Construction Information Systems Australia, Sydney.

76 Roodman, D. & Lenssen, N. (1995) *A Building Revolution: how ecology and health concerns are transforming construction.* World Watch Paper No. 124, March. World Watch Institute, Washington DC, USA. p.22.

77 Willis, A.M. & Tonkin, C. (1998) *Timber in Context: a guide to sustainable use.* Construction Information Systems, Australia, Sydney.

78 Roodman, D. & Lenssen, N. (1995) *A Building Revolution: how ecology and health concerns are transforming construction.* World Watch Paper No. 124, March. World Watch Institute, Washington DC, USA.

79 Edminster, A. (1998) Measuring Forest Impacts of Residential Construction. *Proceedings – Green Building Challenge'98.* 26–28 October. Vancouver, Natural Resources Canada. pp.387–395.

80 Edminster, A. (1998) Measuring Forest Impacts of Residential Construction. *Proceedings – Green Building Challenge'98.* 26–28 October. Vancouver, Natural Resources Canada. pp.387–395.

81 Howard, N. (2000) *Sustainable Construction: the data.* CR 258/99. Building Research Establishment Centre for Sustainable Construction, Watford, UK.

82 For more information on the Forest Stewardship Council – http://www.fscoax.org/.

83 Grey, A. & Hall, A. (1999) *Forest Friendly Building Timbers.* Earth Garden, Trentham, Australia.

84 Lawson, B. (1996) *Building Materials, Energy and the Environment: towards Ecologically sustainable development.* Royal Australian Institute of Architects, Red Hill, Australia.

85 Briffet, C. Mathur, K. & Ofori, G. (1998) Sustainable Development and the Built Environment in Singapore. Proceedings *Building and the Environment in Asia* CIB TG8 National University of Singapore 11–13 February. pp.55–60.

86 State of the Environment Advisory Council (SOEAC) (1996) *Australia: State of the Environment.* CSIRO Publishing, Collingwood, Australia. pp.3–38 as cited in Yencken, D. & Wilkinson (2000).

87 State of the Environment Advisory Council (SOEAC) (1996) *Australia: State of the Environment.* CSIRO Publishing, Collingwood, Australia. pp.3–38 as cited in Yencken, D. & Wilkinson (2000).

88 World Resources Institute (1998) *World Resources Report 1998–1999 Environmental Change and Human Health* Oxford University Press, Oxford, UK.

89 Smith, M., Whitelegg, J. & Williams, N. (1999) *Greening the Built Environment.* Earthscan Publications, London.

90 United Nations Development Program (UNDP), United Nations Environment Program (UNEP), World Bank & World Resources Institute (2000) *A Guide to World Resources 2000–2001: people and ecosystems: the fraying web of life.* World Resources Institute, Washington, DC.

5 SUMMARY: WHAT DO BEES KNOW NOW?

Why interdependency is important to understand

Mapping the life-cycle metabolism of a building provides us with a description of its ecological interdependencies, and the ability to determine the likely ecological impacts of the relationships it has with the natural environment.

One of the key features of an ecosystem is that all communities are dependent on each other for their prosperity, and the functioning of the whole ecosystem is dependent on the prosperity of its communities and the viability of its non-living components. In other words, all elements of an ecosystem are interdependent. Buildings, the built environment and the process of building are all components of ecosystems. Hence the building industry is interdependent with the natural environment and, just like any other element of an ecosystem, our industry's prosperity depends on the health of non-human life and habitats.

The major issues

Life cycles

Buildings usually last a long time. They therefore affect the environment for a long time. What we require of a building today in terms of resource input and what we design its wastes to be, creates relationships with nature that can be more lasting than the structure itself.

Ecologically sustainable building activity relies on nature always having the resources that are required to maintain it, and being ecologically robust enough to assimilate the waste a building produces throughout its entire life. Ecologically sustainable relationships are cyclic: resource paths spiral through a building or product life cycle as outputs are fed back to sustain production.

Metabolism

The environmental impact of building is related to the stocks of resources available for use, and the flows of resources and energy required for operation. Millions of tonnes of material are 'stored' in our built environment along with important human and ecological resources. Yet very little of these resources are reused as a resource for new building development. The metabolisms of built environments are therefore very dependent on inputs of resources and ecological services. Unfortunately much of our energy, water and transport infrastructure generates large quantities of waste and pollution. Waste drives increased resource consumption, while pollution accumulates in receiving environments and reduces their health and, therefore, their capacity to assimilate (clean up and make safe) our emissions.

The major impacts

- Resource depletion
- Physical disruption
- Pollution
- Social and cultural effects

These impacts accumulate in the environment over the life cycle of a building, leading to effects in natural systems that bring changes in environmental conditions. Major concerns are disruptions to global biogeochemical cycles that are leading to phenomena like the enhanced 'greenhouse effect', and environmental pollution. In the next section we explore the interdependencies of building and Earth's natural cycles so we can understand in more detail how building affects nature.

Energy consumption

A more rapid release of carbon into the atmosphere in the form of CO_2 has occurred since the industrial revolution due to the burning of fossil fuels. This has in turn caused a destabilisation of the carbon cycle leading to a build-up of atmospheric CO_2 and an increase in global mean temperature due to an enhanced 'greenhouse effect'.

The building industry is responsible for greenhouse emissions when fossil fuels are consumed to create electricity or fuel for the mining and manufacture of building materials, for their transportation and for on-site construction. The energy required for these processes is called a product's *embodied energy*. Buildings require huge quantities of electrical energy for operation. The source of this energy varies from region to region. Disruption to the carbon cycle occurs predominantly when electricity is generated by burning coal or oil. Energy consumed in transportation of materials, occupants and wastes throughout a building's life cycle is also a contributor to carbon emissions.

Water consumption and pollution

Urban areas increase water run-off and decrease filtration. Drains and channels concentrate water flow and help to transport pollution from streets and building sites into streams, rivers and marine environments. Urban areas in hot climates can increase rates of evaporation of rainfall, due to radiant heat from built form. Phosphates in garden fertiliser and detergents can also be introduced to the water cycle in large quantities, causing increased nutrient loading and stimulating aquatic plant growth in downstream environments. Accelerated plant growth in streams and rivers can reduce water oxygen content, block sunlight, and encourage algal blooms.

Building materials manufacturing and transport

Nitrogenous and sulphurous oxides and carbon dioxide are released into the atmosphere during the production of a range of common building materials. Nitrogen and carbon dioxide are greenhouse gases that are increasing in atmospheric concentration as a result of human industrial activity. This is contributing to the enhanced greenhouse effect. In the atmosphere nitrous oxides can combine with water to form nitric acid and fall as acid rain. The combustion of automotive diesel oil (ADO) also produces nitrogen emissions. A large percentage of all fuel energy used during the construction phase of a building is ADO.

Sulphur oxides (SO_x) are a byproduct of steel, copper, clay brick and cement manufacturing. In large quantities, atmospheric SO_x can combine with water, forming sulphuric acid and fall as acid rain. This phenomenon has in the past devastated environments surrounding mining and manufacturing operations and has damaged building façades.

Biodiversity

The building industry has contributed most to reducing biodiversity through habitat destruction caused by extraction of resources, particularly timber, and changing land use from rural to urban.

It is possible to build in ways that not only reduce impacts of biodiversity, but which can actually restore it. Achieving these aims requires building that:

- minimises resource consumption;
- uses timber from certified sustainable plantations;
- discourages sprawl;
- builds on previously developed sites;
- replaces man-made technology with ecological processes;
- protects and creates habitat; and
- makes provision for growing food.

What is next?

Ecosystems are dynamic and are constantly adapting to changes in bio-geochemical cycles. Ecosystem adaptations, in turn, alter these cycles. Building and urban areas have played a major role in disrupting Earth's natural cycles. The question for us is: can we adapt to the changes we are causing? The next part of the book begins to explore this question by investigating the thermodynamic basis for ecological sustainability.

Remember: everything is connected with everything else.

More information

Life-cycle thinking

United Nations

Life Cycle Assessment: what it is and how to do it (1996) United Nations Environment Program – Industry and Environment, Paris, France.

International Organisation for Standardisation (ISO)

> ISO 14040: Environmental Management – life cycle assessment – principles and framework (1997).

Society for Environmental Toxicology and Chemistry – Europe (SETAC)

> *A Conceptual Framework for Life-Cycle Impact Assessment* (1993) Fava, J., Consoli, F., Denison, R., Dickson, K., Mohin, T. & Vigon, B. (eds).
> *Guidelines for Life-Cycle Assessment: a code of practice* (1993) Consoli, F., Allen, D., Boustead, I, Fava, F, Franklin, W, Jensen, A.A, *et al.* (eds).

Metabolism issues

> Resetting the Compass (2000) Yenken, D. & Wilkinson, D. CSIRO Publishing Collingwood, Australia.
> Newman, P. & Kenworthy, J. (1999) Sustainability and Cities: overcoming automobile dependence. Island Press, New York.
> Rees, W., Testemale, P. & Wackernagel, M. (1996) *Our Ecological Footprint: reducing human impact on the Earth*. New Society Publishers, Gabriola Island, Canada.
> *World Resources 1996–1997: a guide to the global environment* (1996) World Resources Institute, Oxford University Press, New York.
> *A Building Revolution: how ecology and health concerns are transforming construction* (1995) Roodman, D. & Lenssen, N. World Watch Paper No. 124, March. World Watch Institute, Washington DC, USA.

Impacts

> *Architecture and Climate Change* (1992) Szokolay, S.V. Royal Australian Institute of Architects Education Division. University of Queensland, Australia.
> *Building Materials, Energy & the Environment: towards ecologically sustainable development* (1996) Lawson, B. Royal Australian Institute of Architects, Red Hill, Australia.
> *Ecology of Building Materials* (2000) Berge, Bjørn, trans. Filip Henley. Oxford: Architectural Press [original published 1992].
> *Timber in Context: a guide to sustainable use* (1998) Willis, A.M. & Tonkin, C. Construction Information Systems Australia, Sydney.
> *Human Impact on Australia's Beaches* (1996) Legge-Wilkinson, M. Wet Paper Publishing, Ashmore, Queensland, Australia.

Ecosystems

The Web of Life (1997) Capra, F. Harper Collins, London.
Construction Ecology: nature as a basis for green buildings (2002) Kibert, J., Sendzimir, J., & Guy, B. (eds) Spon Press, New York.

Reflection time

Understanding your environmental perspective

At its basic level 'home' is where the heart is. The concept of home, in this case, is the body. It is our own personal construction that shapes our perceptions of the world around us. In this activity you are going to identify, and then understand, the environmental perspectives you are bringing with you into your course. Students come into the course with a whole range of preconceived ideas and ideologies, some informed, others not. In order to provide you with the capacity to approach your professional and personal activities in the built environment with an environmental perspective, it is first necessary for you to understand more about your current perspectives on environmental issues. You also need to be more aware of how you have formed these perspectives.

(a) On a sheet of A3 paper, develop a map of your life so far. Include all the events and people in your life that you feel have shaped your attitudes and opinions, your habits and personality, your hobbies, likes and dislikes. Circle the significant influences on your environmental perspective. Organise your map in chronological order.

Once you have drawn your map take a close look at it. Think about the events and people in your life that have shaped your environmental attitudes. Consider your family history as well. Think back to your grandparents' generation.

- Were your ancestors immigrants or native to your country?
- Were your parents first, second, third or fourth generation immigrants?
- What large events (wars, depressions, revolutions, social movements or global events) shaped their lives? How did your family use the land or relate with nature?
- Which of their values have you absorbed?
- Which of their values have you rejected?
- Who else has influenced your values?
- What is their background?

(b) By considering these questions and referring to your map, discuss with friends how you have formed your current environmental perspective.

Life-cycle thinking

Choose a building material or product. Create a flow diagram of the life-cycle story of that product by following the four-stage process of life-cycle analysis. If you are in a group, have other group members repeat the process on different materials or products.

(a) Present and discuss your results; or
(b) Write a short dramatised story of the product's life cycle, perhaps incorporating some of the people that are involved with, or affected by, the product over time.

Metabolism: home metabolism audit

Houses are connected to nature. Natural resources are required at all stages of the building life cycle. The act of building and then operating a home creates 'flows' of resources from nature to the house. The house then uses these resources and emits byproducts ranging from solid waste to greenhouse emissions back into the environment. This can be thought of as a kind of metabolic system, like our bodies have. As building professionals we need to be aware of the resource flows we create when we build a house and we need to be aware of the environmental implications of these flows. As you will learn, a building's metabolism is a very important indicator of how well it contributes to ecological sustainability.

One of the best ways of learning about problems is discovering them for yourself. The home audit consists of determining what your home requires from the natural environment in order to be constructed and to operate.

Method

Analyse the metabolism of the house that you live in by following these instructions:

(a) Draw a site plan of your home on a sheet of A3 paper, using a scale of 1 : 100 for your house and site.
      Measure the quantities of the following elements of your home, noting the type of material used:
  - structural timber (note the species used and where it comes from);
  - exterior cladding material;
  - interior cladding material;

- roofing material;
- concrete slab (where applicable);
- gross floor area in m²
- gross wall area in m²
- gross window area in m²

In point form, note on your plan the major environmental impacts of these materials in your home.

Indicate in different colours:

- drinking water in;
- 'grey' water outlets;
- 'black' water outlets;
- electrical energy in;
- fuel energy in.

Note on the plan the exact origin of these resources and any destination of outflows.

(b) Using the tables provided, determine the following quantities and note them on the plan:

- annual consumption of drinking water;
- annual consumption of electricity;
- annual CO_2 emissions from electricity consumption (Table 4.1);
- embodied energy of your home (Table 3.4).

(c) Note on your plan any other aspects of your home that might contribute to the following environmental impacts:

- greenhouse gas emissions;
- loss of biodiversity;
- solid waste generation;
- chemical pollution;
- depletion of non-renewable resources.

Note five (5) things that your could do to, or in, your home to best reduce its negative environmental impact and increase its contribution to ecological sustainability.

PART II
BUILDING ECOLOGICAL
SUSTAINABILITY

INTRODUCTION

> 'An organism stays alive in its highly organised state by taking energy from outside itself, from a larger encompassing system, and processing it to produce…a more organised state'.[1]

Gravity is obviously a force that all building professionals understand and incorporate into building design and function without question. If a building is not designed to resist gravity then of course it will fall down. This is directly threatening to the lives of people who are in and around the building when it collapses. Less obvious, however, are the threats to life posed by creating buildings that don't work with laws of thermodynamics.

In this section we will discuss the role of nature in providing energy for our built systems and the role that building can play in using nature's energy in an ecologically sustainable way. We begin by describing how we get our energy in the first place and progress to look at the properties of energy and matter, the laws that describe how they move through human and non-human systems and that govern the design of ecologically sustainable systems. By understanding how energy and matter flow and how nature designs systems to best utilise available resources, we can derive some non-negotiable rules for building that allow us to build with nature, not against it.

In the beginning…

To an individual human being in a large geopolitical system it might seem that the problems confronting human and non-human life on the planet are just too big to deal with. If you feel this way perhaps you will find some solace in the knowledge that the very foundation of life, the engine of creation, the source of the air you breathe is not big at all, in fact it is microscopic, and that the basis of any solution to our global problems is not contained within global economies or politics but in the cellular structure of green leaves.

Within a plant's cells reside organelles called chloroplasts; these contain important proteins, lipids, sugar and nucleic acid molecules which assist the green chlorophyll molecules in capturing the energy of the sun, turning it into sugar and oxygen during photosynthesis. Plants turn sunlight into sugar and oxygen and capture energy into chemical bonds with high potential energy. In this way sunlight is turned into matter with stored potential energy and the atmosphere is charged with oxygen.

This happens during 'light reactions', so called because they happen only when light is being received by a plant's chloroplasts. Earth receives 1.4 kW/m² of energy from the sun. At least half of this energy is reflected by the atmosphere back into space. Most of the rest of the incoming solar energy is absorbed by water and land mass, helping to maintain global temperatures within a range that can support life. A mere 1–2% of incoming solar energy is of a wavelength that can be captured by green plants during photosynthesis.[2]

There is also a set of reactions called the dark reactions that can only happen when light is not being received by a plant's chloroplasts. These reactions, aided by molecules of adenosine triphosphate (ATP), split carbon and atoms from sugars and recombine them with oxygen to form carbon dioxide and water in a process known as cell respiration. The following equations show that during photosynthesis energy is captured, while during respiration energy is released.

Photosynthesis[3]

$$6 \, CO_2 + 6 \, H_2O + light \rightarrow C_6H_{12}O_6 \, (glucose) + 6 \, O_2$$

Cell respiration[4]

$$C_6H_{12}O_6 \, (glucose) + 6 \, O_2 \rightarrow 6 \, CO_2 + 6 \, H_2O + energy$$

Animals like us don't have chlorophyll so we can't participate in photosynthesis. However we do have the equipment for cell respiration, which we use to break down the organic molecules in our food. In the process we consume oxygen and release carbon dioxide.

A lot of energy has been stored in the carbon-based molecules of organic matter, which over many millions of years has become fossil fuel such as coal, natural gas and crude oil. We use this potential energy as fuel to create electricity, and to drive our industry.

As the last chapter pointed out, the way we are using the stored energy sources is creating byproducts that are changing natural cycles, to our detriment. In addition, while nature continues to store energy via photosynthesis, we are using stores at a far greater rate than can be replenished by either nature or ourselves. In essence we have not designed and built ecologically sustainable systems.

This section of the book describes natural laws and phenomena that establish the basis for ecological sustainability. Chapter 6 deals with thermodynamics and provides a basic description of the way energy and matter behave. The conditions described by laws of thermodynamics set the ground rules for ecologically sustainable building. First we understand the rules, then we develop some guiding principles.

Chapter 7 deals with another inevitability of life (and death) and that is change. Conditions on Earth are undergoing constant change. In order to maintain or improve quality of life we need to be able to adapt to change in positive ways. This requires us to accurately perceive the changes that affect us, plan ahead to direct changes in current practice that will positively influence the future, and maintain the resilience of our built and natural systems. The influence of change is pervasive and also sets some ground rules for ecologically sustainable building. As in Chapter 6, first we understand the rules, then we discuss some guiding principles. Chapter 8 provides a summary of these basic ground rules for ecological sustainability.

6 THERMODYNAMICS: UNDERLYING PHYSICAL LAWS

Introduction

French engineer Sadi Carnot (1786–1832) initiated the study of classical thermodynamics in 1824, as he studied the performance of steam engines.[5] He discovered a relationship between the amount of work performed by the engine and the amount of heat produced by the engine as it performed its work. He was trying to work out how to create an engine that would do the maximum amount of work for minimal energy input and he correctly perceived that the key was to reduce the amount of energy being lost as heat. He wanted the engine to perform as much work as possible for as little input of fuel, that is, to be energy efficient.

He found that he could increase the efficiency of the engine if he reduced the friction of the engine's moving parts, thus reducing the amount of energy wasted as heat. In the process he observed that the amount of energy put into the system was equal to the amount of work produced by the engine, less the heat energy produced through friction. In other words, the amount of energy in the fuel used was equal to the energy emitted as heat plus the work produced by the engine.

Carnot's work was refined between 1840 and 1860 by Rudolf Clausius, a Polish mathematician who developed the classical expressions of the first and second laws of thermodynamics.[6] These laws are fundamental to developing principles of ecologically sustainable building.

The laws of thermodynamics explain the way energy and matter flow and transform to create the conditions of function to which all structures and systems comply. While the design of systems *within* buildings, such as mechanical, electrical, and hydraulic systems must apply principles of thermodynamics, it is equally important that they be understood when designing interactions of a building with its *external* environment. BEEs know how to take advantage of the conditions that these laws create by applying their understanding of conservation and efficiency, entropy and surviving designs.

Knowledge of conservation and efficiency

According to Clausius' first law of thermodynamics, energy can neither be created nor destroyed, but instead is transformed into different forms, with the amount of energy in the system after the transformation equalling the amount of energy that existed prior to the transformation.

Under normal circumstances matter can't be destroyed either. The first law implies that energy and matter can't really be consumed, but are instead processed and transformed. The amount of energy and matter existing before the transformation will be equal to the amount existing after the transformation, although in different forms. This means that all matter that makes up the timber, concrete, steel and glass of modern buildings (and the human body, for that matter!) has existed as long as the Earth has in one form or another. The atoms that make up the steel in the Empire State Building in New York might have once been present in the form of a dinosaur, a fern or a jellyfish. It is important to realise that at the atomic level the building industry never makes anything new. Existing matter is just transformed and rearranged by industrial processes to make new forms.

An implication of the first law is that nothing can ever be 'thrown away' because at the atomic level there is no 'away'. BEEs therefore ensure that the material transformations required to construct and maintain a building do not create any forms of matter that can accumulate and become harmful to life. It is ironic that Clausius' 'discovery' that matter is with us forever coincided with the development of a global industrial economy that consumes matter existing in natural forms and, in transforming it, often creates toxic and polluting by-products, the accumulation of which now threatens all life.

BEEs minimise resource consumption and waste and maximise efficiency

Since that time, rates of transformation of finite natural resources from useful to useless forms now far exceed rates at which many can be replaced either by humans or by nature. Scarcity of useful resources like crude oil* is balanced by an increasing abundance of less useful or harmful substances like CO_2 and NO_x. In accordance with the

---

*In March 1997 at the Moscow Summit of 'G8' nations the International Energy Agency (IEA) presented a paper based on the work of geologists Collin Campbell and Jean Laherre, accepting that between 2005 and 2010 supply of conventional oil energy will be outstripped by world demand. Campbell and Laherre assert

first law of thermodynamics, the only way to address this situation is to minimise the transformation of resources into useless forms of matter and energy. This means minimising resource consumption to rates at which resources can be replenished or that allow nature to assimilate byproducts harmlessly, and changing from using resources in ways that cause pollution. The first law of thermodynamics therefore reminds BEEs to be resource efficient, to minimise consumption (dematerialise), and to use resources wherever possible which don't cause pollution and that are easy to regenerate.

The sustainable use of resources requires using renewable resources at a rate at which they can be replenished. Resource efficiency, dematerialisation and recycling are all means to this end. A non-negotiable condition or law for ecologically sustainable building can be crystallised from this discussion of implications of the first law of thermodynamics and stated as:

> **Consume resources no faster than the rate at which they can be replenished.**

Earth's life-support systems are open systems with regard to energy but closed systems in terms of materials. Earth receives new energy from the sun, captures it as heat in its mass, turns it into food, water and oxygen using photosynthesis, and radiates it back into space. Materials, on the other hand, as the first law explains, are never replenished, just transformed. On Earth energy *flows* while materials *circulate*. With the exception of nuclear reactions, the way energy flows is explained by the second law of thermodynamics. BEEs call this kind of knowledge the knowledge of entropy.

Knowledge of entropy

The classical definition of the second law of thermodynamics states that as energy flows through a closed system the energy available to that system for work decreases. The emphasis of the second law is on the quality of energy (also known as *exergy*) in a system. The higher the *energy-quality*, the more capacity energy has to do work. In a closed system the amount of energy available for work (free energy) decreases as it is used, and the amount of energy unavailable for work (bound

that at current production rates world supplies of conventional oil will last for 43 years at best. From: The oil shock to come. Srodes, J. 7 September 1998 *The Age* newspaper Page 15, Melbourne Australia.

energy) increases. The measure of bound or low-quality energy in a system is called *entropy*. Any system which does not receive constant inputs of energy is considered a closed system.

Because energy-quality is constantly degrading, high-quality energy must always be added to a system to keep it organised. As the amount of bound energy increases, less energy is available for work and the system becomes more and more disorganised. Entropy is therefore also a measure of disorganisation.

A piece of coal represents *free* energy because it has high energy-quality and can be used, and is therefore described as a source of low entropy. Burning the coal transforms the free energy into bound heat and light energy and the matter into gas, smoke and ash. If no more coal is added to the fire it will go out as all of the free energy is converted into bound energy, the fire and its aftermath being high entropy.[7,8]

When all energy in a system becomes bound the system is said to have very high entropy and low organisation, a state called thermodynamic equilibrium. The piece of coal, for example, is very organised and is far from equilibrium while the fire, smoke and ash are very disorganised and close to equilibrium. Kitchens and children's bedrooms provide close-to-home examples of systems that become rapidly disorganised without the input of additional energy (to clean them up!).

While the classical definition of the second law explains what happens to energy and organisation in a closed system, it does not sufficiently explain what is going on in nature and the built environment. If it did then you would expect to be living in a very disorganised and unpredictable world. Contrary to the idea of the inevitability of increasing disorder, nature has established processes that increase order despite the constant degradation of energy-quality.[9] Ecosystems have evolved to transform the flow of abundant disorganised 'close to equilibrium' energy from the sun into highly organised 'far from equilibrium' states. Because of this, ecosystems are considered to exist far from thermodynamic equilibrium.

Nature achieves this seemingly counter-thermodynamic feat because it is an open rather than a closed system, constantly receiving new energy from the sun to 'top-up' the energy-quality used to maintain organisation. This solar income is 'stock-abundant but flow limited', while terrestrial energy resources like crude oil and coal are stock limited and (temporarily) flow abundant. As Herman Daly explains:

'We cannot mine the sun to use tomorrow's sunlight today, but
we can mine terrestrial deposits and, in a sense, use up tomorrow's petroleum today.'[10]

Schneider and Kay argue that ecosystems cannot only maintain
order, they can increase their complexity, diversity and stability if
they create systems and relationships that allow them to best use incoming energy. These observations have led to a new description of
the second law of thermodynamics that better explains how energy
flows in open systems. Rather than the emphasis being on increasing
entropy, the restated second law emphasises emerging structures
and helps explain the behaviour of complex, far from equilibrium,
systems.

'The thermodynamic principle which governs the behaviour of
systems is that as they are moved away from equilibrium they
will use all avenues available to counter the applied gradients.
As the applied gradients increase, so does the systems ability to
oppose further movement from equilibrium.'[11]

Observing the structure and processes in natural systems resulting
from the influence of second law phenomena, we notice that consumption of energy-quality is a primary influence. Energy-quality
must therefore be considered concurrently with the quantity of energy used. The second law of thermodynamics provides BEEs with a
very important set of conditions for creating ecologically sustainable
buildings.

BEEs consider the quality of energy being used, not just the quantity

Energy flows through a hierarchy of increasingly organised but less
abundant forms in ecosystems. This hierarchy is known as an energy
pyramid and is organised into *trophic levels* according to how different organisms obtain energy from food for life processes, as shown
in Fig. 6.1.

In any ecosystem in general there will be more plants than herbivores and decomposers, and more primary carnivores than top carnivores. This is because as energy is transformed from one trophic level
to the next its capacity for work is degraded and eventually becomes
unusable heat energy.[12] As energy flows through each trophic level,
the *abundance* of useable energy decreases, while the *concentration* of
high-quality energy increases.

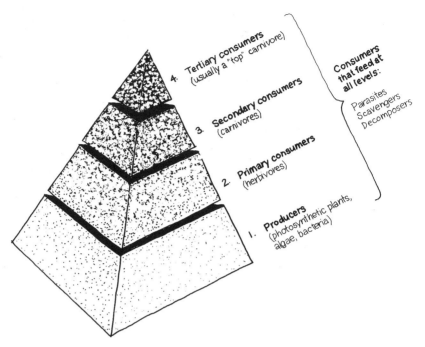

Fig. 6.1 Trophic levels. Each layer of the pyramid represents a different trophic level. Biologists and ecologists identify organisms in an ecosystem by the type of energy they use for their life processes (e.g. producers, herbivores, carnivores, omnivores, parasites, decomposers or scavengers). Organisms can also be identified according to the consumer level they occupy (eg. producers, primary consumers, secondary consumers, tertiary consumers) or by trophic level (1st, 2nd, 3rd, or 4th). Reference: Cunningham, P. & Saigo, B. (1997) *Environmental Science: a global concern.* Fourth Edition. McGraw Hill, New York.

The important lesson for BEEs is that being energy-efficient (doing more with less) is ineffective on its own as a strategy for developing sustainability. BEEs need to consider not just how much energy and material is involved in building, but how effectively the quality of the resource is being utilised. Using electricity generated by a coal-fired power station to run a bar heater might be considered efficient if the room the heater is in is well insulated to minimise heat loss, and the person in the room is wearing a jumper (to reduce required consumption). If, in addition, all the electricity is converted to heat by the bar heater, then the system could be considered efficient and adhering to first law imperatives because the quantity of energy delivered is being fully utilised.[13]

If we look at the same scenario from a second law perspective, with the emphasis on how well the quality of the energy is being used, we

142

would not focus on how much energy the bar heater needs to consume. Instead we would ask why such high-quality energy as electricity is used to do nothing other than heat radiator bars. Electricity is high in energy-quality and can be used to run all sorts of household equipment from computers to refrigerators that provide a range of services. When used to heat an element in a resistance heater, the energy-quality of the electricity is not being put to its highest and best use. The quality of the electrical energy is therefore wasted in a misuse of energy eloquently described by Greenland and Best[14] as 'like using a sledgehammer to drive a panel pin, or a chainsaw to cut butter'.

BEEs create systems that use energy in a large number of small steps rather than a small number of large steps

BEEs interpret the second law of thermodynamics to mean that the more high-quality energy there is, the more potential there is for creating pathways to get the most service from energy as it degrades. The implication of this restatement of the second law is that the more energy a natural system receives, the more complex and diverse it will become as it creates ways of most effectively using all of the energy-quality available to it, a process known as dissipation.

It is easy to understand this concept if we consider that the most complex and diverse ecosystems on Earth occur around the equator, the region of the planet that receives the most solar energy.[15] The concept could perhaps also apply to money. It always seems that the more there is in an economy, the more abundant are the opportunities to spend it!

The systems of relationships in ecosystems that are created to use energy-quality are called dissipative structures. Our buildings are also dissipative structures, but instead of having many ways of using available energy, they often have very few. The important lesson for BEEs is that nature creates ways of using energy in a large number of small stages, making the most appropriate use of both high- and low-quality energy and in the process minimising loss of energy from the system. Buildings are predominantly designed to only use high-quality energy, often in an inefficient way.

The design condition derived from second law phenomena for ecologically sustainable building is therefore to:

Create systems that consume maximum energy-quality.

In order to consume maximum energy-quality, ecosystems require a pattern of relationships between living and non-living components that are organised to keep energy in the system for as long as possible. This pattern is called an autocatalytic feedback loop and is explained by the fourth law of thermodynamics.

Knowledge of surviving designs

Natural systems are solar income-powered and it should be clear by now that the human economy does not rely on solar income alone. However, the difference between our economy and ecosystems is not just in fuel type, but also in organisation. To survive, a system must be designed to use the energy it receives in the most optimum way. Given that energy-quality is constantly lost from a system as it performs its functions, keeping as much energy in the system for as long as possible is very important.

The work of systems-ecologist Howard Odum has shown that sustainable systems are those that recycle the outputs of consumption as resources for production. Sustainable systems therefore don't waste resources they feed production and thereby keep the energy in their systems for as long as possible (see Fig. 6.2).[16]

The example of the tree offered in Chapter 2 is intended to explain how nature operates to *feed* the production of life-supporting goods and services. In this analogy, all of the tree's byproducts directly or indirectly support its life. In other words a tree doesn't produce waste, it produces food for the ecosystem that supports it. Ecologically sustainable buildings should work like trees in an ecosystem.

Ecosystems convert low-quality, highly abundant, solar energy, into high-quality but less abundant forms of energy. In this process the amount of energy available for work decreases. So, to survive, ecosystems must use the high-quality low abundant energy to reinforce the production of low-quality, high abundant energy. This process supports the conversion of low-quality energy sources such as sunlight into high-quality energy stores such as wood.

BEEs keep energy in their building longer using positive feedback loops

In order to maximise the effective and efficient use of energy, surviving ecosystems have therefore evolved recycling or 'positive feedback' loops that use the outcomes of consumption of resources to feed the production of new resources. The consumption of energy therefore becomes a catalyst for the production of more energy. Ecologists call these types of relationships 'autocatalytic feedback designs' and their

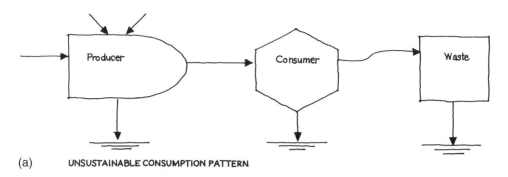

(a)   **UNSUSTAINABLE CONSUMPTION PATTERN**

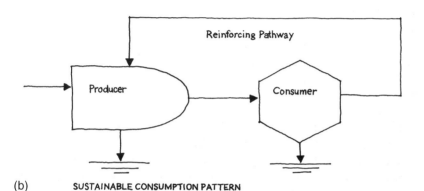

(b)   **SUSTAINABLE CONSUMPTION PATTERN**

Fig. 6.2 Comparison of a surviving design (b) and a non-surviving design (a). Surviving designs are those that use the outputs of consumption as food for production of the resources they need to survive, while minimising waste (b). Non-surviving designs do not reinforce production with the outputs of consumption, instead they accumulate waste (a). Source: Odum, H.T. (1996) *Environmental Accounting: EMERGY and environmental decision making.* John Wiley & Sons, New York. p.281. Fig. 15.1.

effect is to increase the amount of energy available to do work.[17] In other words, they increase power in a system. This is known as the fourth law of thermodynamics and, as expressed by Lotka,[18] it states that: 'Autocatalytic feedback designs develop because they maximise power.' Put simply, surviving designs use 'waste' as 'food' for manufacturing resources, keeping the energy in their systems for as long as possible because that is the most effective way to fully utilise energy-quality.

As we will discuss in Chapter 7, feedback can be positive or negative. Therefore in order for a feedback cycle to lead to sustainability, the outputs of consumption have to be non-toxic to the system and easily reused without the addition of more energy. If the energy required to run feedback is more than the energy in the product produced, then it is not a sustainable process. Surviving systems are, therefore, not only designed to provide pathways for recycling, they create byproducts that are nutrients or raw materials for resource production to ensure positive autocatalytic processes.[19]

The metabolism of our built environment contains some pathways for recycling of materials and energy, however the continuing degradation of ecosystems is an indicator that the flow of energy and materials from human created systems is excessive (see Fig. 6.3). Our systems 'leak' energy and materials into natural systems, causing negative feedback cycles in nature and this is affecting their, and our, ability to survive. The fourth law sets a condition for creating designs that lead to survival and prosperity and this is that built environments must recycle their resources.

In order for recycling to lead to positive autocatalytic feedback, resources available for recycling must be present in sufficient concentrations, in nutritious rather than toxic forms, and must be able to be reprocessed with minimal consumption of additional energy. The design condition derived from fourth law phenomena for ecologically sustainable building is therefore to:

Create and use by-products that are nutrients or raw materials for resource production.

The next section discusses the issues affecting the implementation of these laws to ecologically sustainable building.

Problems with our designs

Comparing the way energy is used in natural systems and the way energy is used in built environments provides a clear picture of which forms of development are ecologically sustainable and which are not. There are two major differences between the way natural systems interact with energy and the way human-designed systems do:

- Nature uses solar energy to create stores of energy. Built environments import energy from Nature but don't replenish stores;
- Nature creates many pathways and feedback loops to use all available energy-quality. Built environments have few pathways and feedback loops and waste available energy-quality.

Fig. 6.3 Energy flow in the Australian building industry. A study of embodied energy flow through the Australian built environment revealed that 45% of all energy entering the through the embodied energy of materials and energy expended in demolishing buildings to access materials for reuse, reprocessing for new materials or reduction is lost in the waste stream. The amount of energy lost from the built environment must be minimised in order to accord with conditions for ecological sustainability. Reference: Tucker, S.N., Salomonsson, G.D. & MacSporran, C. (1994) *The Environmental Impact of Energy Embodied in Construction. Second Report for the Research Institute of Innovative Technology for the Earth*. CSIRO Publishing, Melbourne, Australia. p.7. Fig. 2.1.©1994, CSIRO Australia. Reproduced with permission of CSIRO and the authors.

147

Nature uses solar energy to create stores of energy. Built environments import energy from Nature but don't replenish stores

On Earth energy flows in a hierarchy from diffuse abundant sources to more concentrated and less abundant forms. Nature converts diffuse, abundant solar energy into mass via photosynthesis, thus concentrating energy in highly organised structures. Nature, therefore, is constantly converting sunlight into stores of material with increasing concentrations of energy-quality. The usefulness of an energy source can be described as its 'quality'. Higher quality energy sources are more easily useable than lower quality sources.

Systems ecologist Howard Odum uses the example of how a tree is structured to explain this transformation of energy.[20] A tree has many hundreds of leaves that collect solar energy and transfer it to a smaller number of branches, which in turn transfer the energy to the tree trunk. At each stage work is carried out and a lot of the available energy is degraded and is dissipated as heat. At each stage the quality of energy remaining increases (see Fig. 6.4). A joule of trunk wood energy is considered to be of a higher quality than a joule of sunlight. This is because it is more concentrated and more easily useable by the forest ecosystem and humans.

At an ecosystems scale the same transformations take place. Natural systems develop hierarchies of species to increase the potential of energy that is in the system, an adaptation to the tendency of energy to lose its capacity for work as it moves through a system.[21] Cows, for example, contain higher quality energy than the grass they eat.

In human systems we increase energy-quality predominantly by converting carbon-based substances like coal and oil into electricity. A joule of electrical energy is of a higher quality than a joule of coal energy. However, when we convert fuel energy into kinetic energy, say when we burn petroleum to move our car, we convert high-quality energy into low-quality energy. Similarly, if we use fuel energy for space or water heating we decrease our stock of high-quality energy. Perhaps the greatest degradation of energy-quality occurs in buildings when electricity is used for space heating, a direct conversion of very high quality energy to the lowest quality heat energy. We therefore increase our reliance on diminishing stores of energy resources from natural systems, which is not a sustainable strategy.

This situation creates the imperative for resource efficiency. Because we rely on stores of energy we can't replenish, the less stores we use to do useful work, the longer they will last. Note, however, that being efficient with resources without replenishing them still decreases stores of high-quality energy, just at a reduced rate. Moving

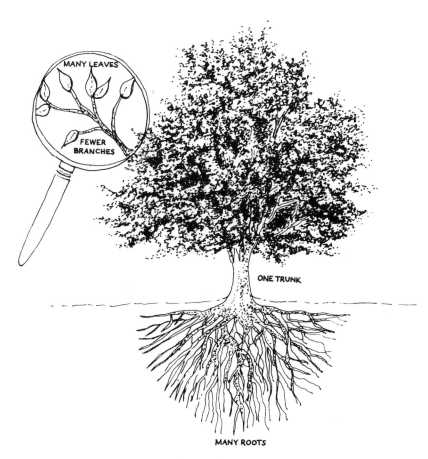

Fig. 6.4 The tree's structure is a manifestation of the energy hierarchy. The design of a tree concentrates diffuse solar energy through many leaves into concentrated storage in the trunk. Diffuse nutrients in the soil are also collected and concentrated through the tree's root system into the trunk. Energy is lost at each transition of energy from leaf to stem to branch to trunk and from root to trunk but the quality of energy in each new structure is greater . A joule of trunk wood energy has a higher quality than a joule of sunlight because it has a greater capacity to do work. Reference: Odum, H. (2002) Material circulation, energy hierarchy, and building construction. In: Kibert, J., Sendzimir, J. & Guy, B. (2002) *Construction Ecology: Nature as the basis for green buildings.* Spon Press, New York. pp.35–71.

toward a built environment that is ecologically sustainable therefore requires more than just energy and resource efficiency.

The challenge for BEEs is to use our fossil fuels predominantly to create the technology we need to capture low-quality energy sources such as solar energy. Consumption of fossil fuel to run a construction process, for example, might be justifiable if the building created makes direct use of solar energy for its operation. Addressing this challenge leads to the second major difference between the energy use of natural and human designed systems.

Nature eliminates waste of available energy-quality using feedback mechanisms. Built environments produce waste by using materials that are not used again, because of poorly structured economic systems, and poor building design.

In order to use all available energy-quality, natural systems develop feedback mechanisms, a phenomenon described as the fourth law of thermodynamics.[22] Through feedback, natural systems:

(1) use the energy-quality in byproducts of one process as feedstock for producing more resources for the system; and
(2) use materials that the system can recycle with minimal additional energy required.

Natural systems therefore evolve to eliminate the waste of potential energy. Energy that is too degraded to be able to be used dissipates into the atmosphere as heat.

In the building industry energy is wasted. While some building industries have developed feedback mechanisms for reusing, reconditioning and recycling, the extent to which the useful potential of existing buildings and materials is captured is by no means complete. While the materials that are in the built environment now represent stores of finite resources with the potential for reuse, in many cases reuse, refurbishment and recycling require the input of large quantities of finite fuel resources.[23] (See Fig. 6.5.) So, while they represent the feedback (or retention) of material utility in built systems, their effect on natural systems is still to increase disorder. Building industries therefore need to evolve feedback mechanisms that require little energy input to operate and which amplify the utility of the systems they are contributing to.

This chapter has been concerned with the universality (unchanging nature) of physical laws of the universe. In the next chapter we consider the concept of change over time, and how the process of change within the physical world impacts on the built environment.

Remember: There is no such place as AWAY. Minimise quantity and maximise quality. Positive feedback is the key.

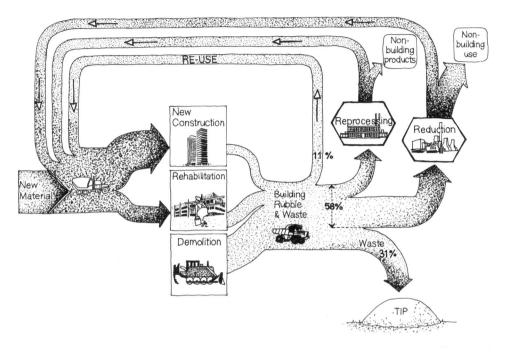

Fig. 6.5 Material flow through the Australian construction industry. The amount of material wasted is an indication of the ecological sustainability of a building industry. Material wastage occurs during new construction, rehabilitation (refurbishment) and demolition of buildings. As this figure shows, approximately 11% of wasted materials are reused directly, while 58% of waste materials are either reprocessed or reduced. Approximately 31% of wasted materials from construction and demolition ends up being dumped. Closing material cycles and minimising material waste are essential for ecologically sustainable building. Reference: Tucker, S.N., Salomonsson, G.D. & MacSporran, C. (1994) *The Environmental Impact of Energy Embodied in Construction.* Second Report for the Research Institute of Innovative Technology for the Earth. CSIRO Publishing, Melbourne, Australia. p.15. Fig. 3.2. ©1994, CSIRO Australia. Reproduced with permission of CSIRO and the authors.

References

1 Schneider, E. & Kay, J. (1994) Life as a manifestation of the second law of thermodynamics. *Mathematical Computer Modelling* **19** (6-8) Elsevier Science, Amsterdam, Netherlands.

2 Cunningham, P. & Saigo, B. (1997) *Environmental Science: a global concern.* Fourth Edition. McGraw Hill, New York. p.51.

3 Cunningham, P. & Saigo, B. (1997) *Environmental Science: a global concern.* Fourth Edition. McGraw Hill, New York.

4 Manahan, S. (1997) *Environmental Science and Technology.* Lewis Publishers, Boca Raton, Florida, USA.

5 O'Connor, J. & Robertson, F. (1998) *Sadi Nicolas Léonard Carnot* On-line article published by the School of Mathematics and Statistics, University of St. Andrews, Scotland. http://www-history.mcs.st-and.ac.U.K./~history/Mathematicians/Carnot_Sadi.html.

6 Schneider, E. & Kay, J. (1994) Life as a manifestation of the second law of thermodynamics. *Mathematical Computer Modelling* **19** (6-8) Elsevier Science, Amsterdam, Netherlands.

7 Georgescu-Roegen, N. (1976) The entropy law and the economic problem. In: *Energy and Economic Myths: institutional and analytical economic essays.* Pergamon Press, New York. p.54.

8 Cunningham, P. & Saigo, B. (1997) *Environmental Science: a global concern.* Fourth Edition. McGraw Hill, New York. p.56.

9 Schneider, E. & Kay, J. (1994) Life as a manifestation of the second law of thermodynamics. *Mathematical Computer Modelling* **19** (6-8) Elsevier Science, Amsterdam, Netherlands.

10 Daly, H. (1996) *Beyond Growth.* Beacon Press, Boston. p.30.

11 Schneider, E. & Kay, J. (1994) Life as a manifestation of the second law of thermodynamics. *Mathematical Computer Modelling* **19** (6-8) Elsevier Science, Amsterdam, Netherlands.

12 Odum, H. (1957) Trophic structure and productivity of Silver Springs, Florida. *Ecological Monographs,* Vol 27, Ecological Society of America, Washington, DC, USA. pp.55–112.

13 Kay, J. (2002) On complexity theory, exergy and industrial ecology. In: Kibert, J., Sendzimir, J. & Guy, B. (eds) *Construction Ecology: nature as the basis for green buildings.* Spon Press, New York. pp.72–107.

14 Greenland, G. & Best, R. (2001) Energy-quality. In: Langston, C. & Ding, G. (eds) *Sustainable Practices in the Built Environment.* Second edition. Butterworth Heinemann, Oxford, UK. p.169.

15 Schneider, E. & Kay, J. (1994) Life as a manifestation of the second law of thermodynamics. *Mathematical Computer Modelling* **19** (6-8) Elsevier Science, Amsterdam, Netherlands.

16 Odum, H.T. (1996) Environmental Accounting: EMERGY and environmental decision making. John Wiley & Sons, New York. p.281. Fig. 15.1.

17 Odum, H. (2002) Material circulation, energy hierarchy, and building construction. In: Kibert, J., Sendzimir, J. & Guy, B. (eds) *Construction Ecology: Nature as the basis for green buildings.* Spon Press, New York. pp.35–71.

18 Lotka, A. (1922) Contribution to the energetics of evolution. *Proceedings of the National Academy of Sciences,* Stanford University Press, San Francisco, USA. pp.148–154.

19 Odum, H.T. (1996) Environmental Accounting: EMERGY and environmental decision making. John Wiley & Sons, New York.

20 Odum, H. (2002) Material circulation, energy hierarchy, and building construction. In: Kibert, J., Sendzimir, J. & Guy, B. (eds) *Construction Ecology: Nature as the basis for green buildings.* Spon Press, New York. p.38.

21 Odum, H. (2002) Material circulation, energy hierarchy, and building construction. In: Kibert, J., Sendzimir, J. & Guy, B. (eds) *Construction Ecology: Nature as the basis for green buildings.* Spon Press, New York. p.38.

22 Lotka, A. (1922) Contribution to the energetics of evolution. *Proceedings of the National Academy of Sciences,* Stanford University Press, San Francisco, USA.

23 Tucker, S.N. Salomonsson, G.D. & MacSporran, C. (1994) *The Environmental Impact of Energy Embodied in Construction.* Second Report for the Research Institute of Innovative Technology for the Earth. CSIRO Division of Building, Construction and Engineering. Highette, Australia.

7 CHANGE: ECOLOGICAL SUSTAINABILITY THROUGH TIME

Introduction

> 'It really boils down to this, all life is interrelated. We are caught in an inescapable network of mutuality, tied into a single garment of destiny. What affects one directly, affects all indirectly.'
>
> Rev. Martin Luther King[1]

Our world is changing. More people have been born on Earth in the last fifty years than in the preceding four million years of human existence. The world's population has more than doubled since 1950, as has average personal income. The combination of these two factors has seen an increase in the size of the world's industrialised economy, which in turn has seen a commensurate rise in the rates of resource consumption and waste production. The total growth in output of goods and services between 1984 and 1994, for example, was four trillion US dollars. This is more economic growth than from the beginning of civilisation to 1950.[2]

The number of people moving to cities is also increasing. The global average growth in city populations is 3.1% or approximately 186 million people per year. This urban growth is occurring disproportionately in developing countries that can least afford to provide even the most basic infrastructure for the new arrivals.[3]

The world is getting warmer. The global average surface temperature has increased 0.6 degrees Celsius over the 20th century, likely to be the largest increase in 1000 years. The 1990s was the warmest decade recorded since 1861. It is thought that this will lead to sea level rises of 2 cm per decade – enough to inundate many low-lying countries like Tuvalu, one of a number of small Pacific Island nations with an elevation of less than four metres above sea level, and alter local climate patterns.[4]

In the time it will take to read this chapter approximately 140 hectares of Indonesian rainforest will have disappeared, about 9930 people will be born,[5] and 44 million litres of sewage effluent will have entered the ocean on the coast of the Australian State of Victoria.[6]

These conditions pose a serious threat to the continued development of human society by placing unprecedented pressure on the assimilative and adaptive capacity of the natural environment. They indicate the imperative to change the kinds of human activity that exacerbate such conditions, and compel us to proactively identify and implement new methods of development.

While it is change that is being called for, the popular rhetoric calls for sustainability, a concept described by the Oxford dictionary as meaning 'To keep going indefinitely'.[7] We obviously cannot keep following existing development patterns, so how do we change and what do we change in order to build in an ecologically sustainable way? We can learn from the BEEs. In this chapter we therefore ask:

- How do BEEs determine what kind of change they need to initiate?
- In a dynamic world is the best way to contribute to ecological sustainability to make buildings that resist change or buildings that absorb change?
- What kind of change should be initiated in order to change current building practices so they contribute to ecological sustainability?

We begin by looking at the factors that influence our ability to affect change. Changes occurring in nature are a result of many complex feedback loops. The relationships between cause and effect are therefore non-linear. Unfortunately, many decision-makers have been trained to think of linear cause-and-effect relationships and this gets in the way of their noticing or even considering environmental consequences. We begin this chapter, therefore, in the mind of the BEE and discuss the nature of change and what influences our perception of it.

BEEs learn from experience and from observing nature. In this way they develop an understanding of how to build in ways that deal with external and unexpected change and avoid building obsolescence. As we will discover, this is not just a matter of choosing the right materials or building design.

Avoiding obsolescence is about being able to create buildings that continue to meet people's needs. BEEs facilitate this by designing a system of relationships between stakeholders that allows for feedback and learning, the integration of values, and a diversity of social and economic perspectives and strategies for harmonising with emerging environmental conditions. The concepts of ecologically sustainable development and sustainable construction are important movements that are attempting to provide approaches to realising these aims. We therefore look at how BEEs' activities are contributing to these broader movements of social change.

When BEEs build they intend to meet the needs of their client, and to improve the health of natural systems. Their intention is to create an ecologically sustainable building. However, whether a building is or is not ecologically sustainable depends on the resilience of the ecosystems it affects and how the building itself can adapt to changing environmental conditions. We therefore need to understand the relationships between BEEs' intentions and the surprises nature has in store.

There are essentially two kinds of change: changes that we plan (intentions) and changes that we don't (surprises). Intention comes from within our bodies – it is a product of our perception of the suitability of a current situation. Surprise occurs when situations arise that we do not expect – it is to these we must adapt. This observation generates two questions that BEEs tackle when advancing ecologically sustainable building. They are:

- What kind of change should be planned in order to change current building practice so that it contributes to ecological sustainability?
- How should buildings cope with uncertain futures?

Intention

In answering the first question we must develop an understanding of what aspects of current practice need changing. What we identify as needing change depends on what we perceive is leading to environmental problems. Our ability to make the right decisions about ecologically sustainable building therefore depends on how we notice environmental change, and what we perceive the causes to be. Correctly identifying causes of environmental change depends on whether we think linearly or non-linearly about cause and effect. We then decide on a course of action.

BEEs are skilled in planning changes that contribute to ecologically sustainable building in a dynamic environment. This skill requires:

- accurate perception of change;
- understanding non-linear cause and effect;
- receiving appropriate feedback; and
- learning by reflecting.

Perceiving change

The ability of organisms to perceive change and adapt is a key life-supporting feature of our planet. The perception and adaptation of Earth's organisms to changes in levels of solar radiation, for example, have kept Earth's average surface temperature within a range that has supported life for about four billion years despite a 25% increase in the heat of the sun over the same period.[8] Even at the basic physical level, an electric current will not flow unless there is a positive and a negative charge present. It is clear that life is not only conditional on the ability to perceive and adapt to change, but that change is itself necessary for life to exist.

In order to create buildings that effect positive change in the environment, BEEs must understand something of how environmental changes occur. Having a basic understanding of the nature of change empowers BEEs to make decisions that enhance positive change and avoid negative change. There are two basic conditions that need to be understood:

- that organisms perceive change by noticing difference;[9]
- that non-linear relationships exist between elements of an ecosystem that amplify change through feedback.[10,11]

The perception of difference

'I have discovered that one of the most important economic trends is that they are too slow in their motion to be visible to humans…Humans do not get out of the way of that which they cannot see moving. As with the electromagnetic spectrum, most of the frequencies and motions of Universe, are ultra or infra to man's sensory tunability.'[12]

Imagine that all the words on this piece of paper were the same colour as the background. If   were   would   impossible   read. If, on the other hand, the words on the page were only a slightly different colour to the background, it might be possible to read them but we would have to really concentrate. The words would become easier

157

to read as the colour of the print contrasted more with the colour of the paper. All life in this way perceives change by noticing a difference between the background or 'ambient' conditions and some other sensory stimuli. According to Senge[13] there are three features of our environment that can increase the difficulty with which difference and thus change can be detected. The first is the sensitivity of the sensors that we use to receive information about our surrounding environment; the second is the uniformity of our background conditions or the level of 'noise' present. The third factor affecting our ability to notice difference is how quickly a difference occurs or the 'rate of change' being experienced.

Sensitivity of our senses

We use our senses of touch, taste, sight, hearing and smell to experience differences in our environment. Each sense has a threshold of difference beyond which it cannot perceive change. With healthy hearing we can hear a pin drop if we are concentrating, but cannot hear a dog-whistle. In this case the frequency of the dog-whistle is beyond our sensory threshold. Assuming that blowing the whistle has not caused any changes in the environment because we have not noticed any change would be incorrect. We would be completely unaware that a change in our environment (like a stampede of German shepherds!) had occurred until we encountered the stampeding dogs. In a similar way we do not directly perceive increases in atmospheric concentrations of CO_2 because it is a colourless, odourless gas that is already present in the atmosphere, and thus already an aspect of 'ambient' atmospheric conditions. We have, however, noticed changes in weather patterns and are drawing the conclusion that these changes might be associated with increased concentrations of atmospheric CO_2.

Noise

Our senses also have a threshold of perception related to the amount of information they are receiving from the environment at a particular time. Too much information can create as many problems for perception as too little. When you are reading, for example, your eyes perceive the characters on the paper, and from experience you know that the symbols on the paper and the spaces between them are all more or less relevant to the message you are receiving. But a b c if a b c other a b c characters a b c are a b c present a b c you a b c have a b c to a b c decide a b c which a b c letters a b c and a b c spaces a b c are a b c relevant a b c to a b c the a b c message a b c and a b c which a b c are a b c not. The message therefore becomes harder to read. In other words the greater the level of ambient noise, the harder perception becomes.

The urban environment is itself 'noise' getting in the way of the message of the need for ecological sustainability. Many of us spend our lives in cities; we go to the store to buy packaged food – with money from ATMs – and later, after a bit of internal processing, we flush it down the toilet (adding about 6 litres of fresh water to it). Our urban environment – the houses we live in, the roads we drive on and the building we are in today – have all been imposed on a landscape of estuaries and river deltas, basalt plain, river valley, grass and wood-lands that we can no longer see, smell, touch or taste. Big city life, and the nature and shape of our urban environment, are a human con-struct; much of what we encounter is therefore human artefact. This virtuality keeps us from realising, not only that we are connected with nature, but that we *are* nature and that our fate is shared.

A recent study of young Singaporeans' perceptions of nature, for example, revealed that growing up in a highly urbanised environ-ment in which contact with nature is limited, among other factors, has contributed to a general lack of interest and affinity with nature. This attitude is reported to lead to a predisposition for them to adopt '...the rationality of the State in privileging development priorities above conservation imperatives'.[14]

Rates of change

Another environmental condition that affects our perception of dif-ference is the rate at which the difference occurs. Generally, gradual change is far harder for us to perceive than rapid change. This is a par-ticularly important issue for building practitioners because it takes time to build a building, which in turn generally has a long service life. The environmental changes likely to be caused by a building project over its life cycle are therefore likely to be gradual and long-term. That gradual change is harder to perceive is also a major prob-lem for scientists and activists trying to alert those with appropriate authority to safeguard ecological health.

In ecosystems, rapid change is often preceded by very gradual changes in the stress the system is experiencing.[15,16] There are many examples of major ecological disasters that were building gradually but were not noticed until the ecosystem collapsed, a catastrophic change that can happen rapidly.

For example, the trout fishery in Lake Michigan, USA, collapsed in the early 1950s with fish numbers plummeting unexpectedly to near extinction in less than five years. The four years preceding the collapse in fish numbers saw average catch numbers of the trout increase. The sudden decline of the trout was thought to have been brought about by a combination of intense fishing pressure, introduction of foreign predator fish species via the construction of the Welland Canal,[17] and changes in water chemistry.[18] Similar pressures have caused a rapid decline in the 350 predominant species of fish in Africa's Lake Victoria after the introduction of Nile perch in about 1976. Lake Victoria is the world's largest tropical lake, and site of one of its greatest freshwater ecological disasters.[19]

The moral of these stories for BEEs is that decisions about building need to be made based on a consideration of the entire system within which a building interacts. And that just because we don't notice change occurring in the system doesn't mean it isn't happening.

We often concentrate on the parts of a building's environmental influence – the environmental impact of the materials, the remote effects of energy consumption, or the local impact of overshadowing – while our design tools help us conduct experiments on parts of the system we affect. Thermal modelling tools for energy consumption, life-cycle tools for materials, design tools for shading are examples. Traditionally it is up to building designers and engineers to synthesise what they discover about the parts into some understanding of a building's effect on the whole system. This might present a barrier to more widespread adoption of ecological decision-making.

Canadian research suggests that when implementation of environmental assessment and practice occurs, it is only applied partially and selectively.[20] Ecosystems are dynamic, and effects are often too long-term or spatially large to be understood by anything other than a whole systems view.[21]

One of the reasons that gradual change is harder to perceive is because there are delays in the reception of feedback between the cause of a change and the effect we experience.

Feedback

BEEs are systems thinkers, and always begin any decision-making process with a big-picture view of the whole system they are going to affect. The pattern of organisation of an ecosystem is a network pattern within which feedback loops exist to circulate energy, matter and information through systems.[22] The relationships between living and non-living components of an ecosystem are, therefore, non-linear.[23]

160

The relationship between building and nature is therefore also a non-linear relationship. Causes lead to effects, which lead to more causes and other effects. In this way, effects can 'snowball', and phenomena like vicious cycles and self-fulfilling prophecies can occur.[24]

Thinking in terms of linear cause and effect makes it hard, if not impossible, to perceive all of the changes taking place as a result of our actions. Because environmental impacts caused by buildings result from non-linear cause and effect, we may not perceive that our actions are leading to changes in the environment at all. Looking at the problem the other way around, if a change is noticed in an environment, thinking linearly may never lead to understanding possible causes of the noticed effect. Because nature works in systems and not straight lines, BEEs must think in systems to correctly perceive the cause and effects of environmental changes. Consider the following actual cases. What would the BEE's answers be?

- What is the link between salmon populations in Puget Sound and construction in Seattle?
- What is the link between mudslides and floods in the Kalang Valley, Malaysia and construction sites in Kuala Lumpur?

(Turn to Chapter 8, activities, p. 197 for the answers.)

Delays in receiving feedback

Feedback loops provide ways of channelling information through networks of living organisms, within organisms and between organisms and their surrounding environment. Perception of change and feedback are therefore interrelated. Laws of physics best explain why every action we perform has an effect on the environment, however the type of change that we perceive we have caused depends on the feedback we receive through our senses from the environment and the meaning our mind makes about what has happened. In many cases the actions we perform cause changes that are gradual and therefore we do not perceive we have had any effect on the system we are trying to change. The feedback from cause to effect is, in essence, delayed.

To understand the effects of delayed feedback, imagine that you live in a house with dodgy plumbing and you are about to use the shower. First you turn the hot water on and hot water immediately starts to flow. Next you turn on the cold water but the shower temperature doesn't moderate at all so you turn the cold on a little more, the shower temperature begins to fall to an acceptable level and you get in. Everything is fine for a minute or so and then you notice the water temperature beginning to drop and before you can get to the cold tap to turn it down, the shower goes completely cold. From the side of the shower recess you turn the cold down and the temperature becomes reasonable again, and then after another minute begins to get very hot.

You are forced to keep adjusting the temperature of the water the entire time you are in the shower because there has been a delay in your receiving feedback about the difference in temperature that your adjustments of the taps have made. You therefore have to rely on your experience to predict what the effects of your actions are likely to be.

Changes in natural systems may take many years to become noticeable. Perhaps it is because of the lack of immediate feedback about the environmental effects of disruptions to Earth's biogeochemical cycles discussed earlier and not just because of greed, that many captains of industry still steer economic courses toward greater consumption of resources, despite the mounting scientific evidence that this course is damaging ecosystems. There has also been widespread scientific concern that it will take some time to turn the ship of state toward an ecologically sustainable course and that the magnitude of ecosystems damage will continue to increase for some time after a new direction has been established.

The problem of delays in feedback about the effects of actions taken is a common experience for BEEs and is one of the difficulties faced when developing ecologically sustainable buildings. Environmental consequences of decisions made about building design and orientation, for example, may be made with the intention of reducing the life-cycle energy consumption of a building. However, like the unfortunate shower taker, the designer must rely on experience to anticipate the environmental effects of the actions taken. Drawing on data from many previous projects or using sophisticated modelling to choose building materials, plan shape and orientation, the designer can anticipate which configuration will have the greatest positive effect. However no feedback about the actual effect of these decisions on a particular project will be received until the building has been operating for some time.

People taking action to improve environmental conditions very often receive no feedback at all about the effects of their actions, principally because changes in natural environments can take an extremely long time to be noticeable. But their actions do make a difference.

Sometimes the feedback they receive is negative. For example, a popular bumper-sticker in logging towns in Australia reads 'Do something for the environment – Bulldoze a greenie', reflecting the concern loggers have for the viability of their industry and suggesting that the contribution of conservationists opposed to logging would be more valuable to the environment if they were killed and used as fertiliser rather than trying to stop the destruction of forests.

BEEs use building projects as learning opportunities

In order to create ecologically sustainable buildings BEEs have to think not only about the building, they have to think 'outside the box', and consider ecosystems and how they might be affected by decisions made. While thinking outside the box is certainly a necessary skill for BEEs, it is prudent not to ignore what the box might be able to teach us about our relationships with the 'outside world'.

The following are examples adapted from some ideas by Australian Architect John Gelder[25] of ways building can teach people about their relationships with nature:

- *Learn-scaping* – providing a botanical description of plants including indigenous locations, associated animals and insects. Involve clients in propagation and planting.
- *Provide recycling areas, weather stations, solar stations* – provide charts or visual indicators of changes in rainfall, temperature, solid waste production and amount recycled.
- *Exposing building services and structure* – exposing links to material supply chains and indoor/outdoor climate control.
- *Recycled material biographical labels* – tell the story of the origin of recycled materials used. If it is a school, encourage students to research and write each material's biography.
- *Nametags* – for building parts and functions.
- *Low-automation or even manually operated autonomous systems* – occupant operated and maintained technology.
- *Exposed meters* – for energy sources showing how much energy is flowing to (or from the building in the case of grid interactive systems).

As Fig. 7.1 shows, the occupants of this UK house receive immediate feedback on resource consumption because the water and electricity meters are displayed in the kitchen.

It is also important to provide feedback during the construction of an ecologically sustainable building and this can be very important in motivating site staff to participate in environmental management activities.[26,27] In 1998 an Australian study of construction sites operating environmental management programmes[28] identified measures

(a)

(b)

Fig. 7.1 Feedback designed into the BEDZED housing development in Sutton, UK. (a) and (b) Utility meters installed in the kitchen provide immediate feedback on resource consumption. Photographs courtesy of Bill Dunster Architects.

for providing appropriate feedback to construction workers on their contribution to environmental performance objectives. These measures include:

- visible information about the environmental policy and objectives;
- site inductions that include descriptions of environmental plan operation and objectives;
- posting goals and objectives where workers can read them, making sure to cater for a multilingual workforce;
- monitoring performance;
- goal-setting by the workforce – set challenging but achievable goals;

Fig. 7.2 'Waste less' sign. Feedback for workers on their performance helps achieve the best results on-site. These signs were erected during construction of the Docklands Stadium project in Melbourne Australia. Every two weeks the quantity of material sent to land fill was compared with the quantity sent off-site for recycling. The two figures were compared in relation to the quantity of material, in this case timber, being used on-site during that period. Increased recycling rates got a thumbs up, decrease got a thumbs down. Supervisors were then asked to set targets for recycling for the coming two-week period. The target is shown as a dashed line. Actual performance in relation to the target was then plotted to provide feedback on recycling performance. Reference: Lingard, H., Smithers, G. & Graham, P. (2001) Improving solid waste reduction and recycling performance using goal setting and feedback. *Construction Management and Economics* **19** 809–817.

- charts and signs that illustrate the achieved performance of the workforce (see Fig. 7.2) – show whether performance has improved or declined;
- reward – when goals are achieved;
- education – workers need to know why the environmental programme is important. They must understand how they are making a positive difference;
- lead by example – daily operations of site administration must be consistent with the environmental programme goals and must be visible.

Now we have discussed how to perceive and direct change, what happens when something unexpected happens? In addition to fostering positive change, ecologically sustainable building must be able to deal with surprise.

Surprise

Ecosystems exhibit a wide range of variation in complexity and rates of change. Typical properties of ecosystems include complexity, self-organisation, positive and negative feedback, emergence, surprise, inherent uncertainty and limited predictability.[29] Making predictions about our likely effect on ecosystems and the effects of ecosystem change on us is difficult in this context. Decision-making is further complicated by the necessity for trade-offs between competing human social, economic and environmental values.

Under these circumstances it is tempting to simplify problems, squeezing them to fit within the predictive capacity of our decision-making tools so that we can generate an answer that is as 'correct' as possible. Unfortunately, the indicators we can measure might not be the indicators that are important for ecosystem health.

In Sydney, Australia, for example, very sophisticated modelling for predicting energy consumption in a building can be conducted, but little attempt is made at predicting the ecological impacts of additional stormwater or sewage discharge from buildings. Botany Bay, the site of Captain Cook's first landfall in Australia, receives urban run-off discharged into the Alexandra Canal and the George and Cook's Rivers. The 1998 'State of the Bay' study report concluded that Botany Bay is overloaded with nitrogen because of these discharges.[30] From an ecosystems perspective the water emissions of a building in the Botany Bay catchment might be far more important to ecological sustainability than energy consumption on a local scale.

In order to be able to build *with* nature BEEs apply modes of decision-making and create buildings that deal with uncertainty about future conditions, rather than being restricted to approaches that are limited in scope to what can be predicted. BEEs don't know how ecological, social, economic or cultural conditions might change in the future, so in order to deal with unplanned change they take steps to protect the building – not from change – but from obsolescence.

BEEs might try to achieve this by creating buildings that are incredibly durable, designed and constructed to stand the test of time. Like boulders in a stream, they resist the current of change. But even the sturdiest rocks are eventually worn down. Another approach is to create resilient buildings. Resilient buildings are like reeds in the stream, which move to absorb the current and, as long as the current is not too strong, will not be destroyed by it. These two perspectives mirror perspectives of ecosystems.

One view sees ecosystems as existing in an equilibrium steady state, the structure and function of which are altered when disturbed, but return rapidly to an equilibrium state once the disturbance is over.[31] In this view, the faster an ecosystem can recover from the effects of the disturbance, the more stable it is. Buildings tend to be designed as steady state systems, durable and resistant to change. Unfortunately, as they age, they can become more expensive to maintain.[32]

The other view is that of ecosystems as dynamic systems that can maintain their system of relationships and function in the face of the disturbance.[33] According to this view, ecosystems display the quality of *resilience*. Resilience is the ability of an ecosystem to absorb change and disturbance while still continuing to maintain relationships and function.[34] The measure of resilience is the magnitude of disturbance or change that a system can cope with. It has been shown that resilience in ecosystems increases with the diversity of species with overlapping functions at different scales within an ecosystem.[35] There are, however, minimum and maximum thresholds beyond which an ecosystem, built environment and building will be damaged. As a system is pushed to either its upper or lower thresholds it becomes stressed.

Like ecosystems, built systems are designed to cope with changes within certain tolerance limits. Unlike ecosystems, however, buildings that are not either adequately stable or resilient to deal with change don't collapse and die, they become obsolete. In the built environment a measure of both stability and resilience is necessary.

BEEs avoid building obsolescence

Buildings become obsolete for a range of reasons, many of which have less to do with structural wear and tear than with changes in human taste and investment environments. Essentially buildings become obsolete when they no longer fulfil human needs. Buildings can become obsolete due to:

- structural obsolescence;
- economic perspectives;
- user perspectives; or
- social perspectives.[36]

Structural obsolescence

The designed life of most buildings these days is between 50 and 100 years, yet the components of a building will wear at different rates. Building components such as carpet tiles and electrical components, for example, wear out more quickly than structural components of a building. Research into how buildings change over time found that exteriors are changed approximately every 20 years, wiring, plumbing and climate control systems are replaced every 7–15 years, while floor plans last as little as 3 years before they are changed.[37] The different turnover rates of building components are shown in Fig. 7.3 and have been described as a building's 'shearing layers of change'.[38]

BEEs design buildings in ways that allow easy access to so-called fast cycling materials without the need to destroy slower cycling materials (having to destroy ceiling tiles to replace air conditioning registers or hacking masonry to replace plumbing fittings, for example). If they are not designed in this way then the motivation and financial cost to maintain a building can be quickly exhausted and it can be declared obsolete long before it becomes structurally unsound.[39] Maintenance planning is therefore essential for realising the environmental benefits of designing with shearing layers of change in mind.

Economic, user and social perspectives

The length of time a building is designed to last structurally is called its designed life, while the length of time it remains economically viable in the property market is called its service life. Many structurally sound buildings are demolished because they are perceived as no longer viable in the market. When market demand for the services a building is designed to offer falls, a building may be demolished well before its structural use-by date.

Most office buildings in Hong Kong's city centre, for example, are less than 20 years old despite being required to have a 'design life' of 50 years.[40] While the structure of a commercial building may reach the end of its service life within 20 years, its interiors are likely to be

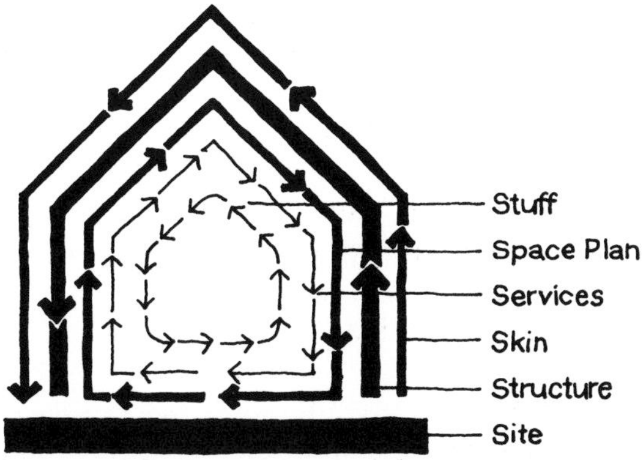

Fig. 7.3 Shearing layers of change. The arrows in this diagram indicate the rate at which different parts of a building change. 'Stuff' or consumables within the building like light bulbs, office supplies, food and furniture spend little time in the building. Designers need to consider how to cater for the inflow and outflow of these materials in environmentally friendly ways. The 'services' and 'space plan' associated with interior fit-out change more often. The building 'structure' changes least but provision needs to be made for more frequent changes in a building's 'skin' or cladding. Designing for different rates of change so that valuable materials or components are not destroyed during adaptation is a key aspect of an ecologically sustainable building design. From the original diagram by Donald Ryan in Brand, S. (1994) *How Buildings Learn: What happens after they're built.* Used by permission of Viking Penguin, a division of Penguin Putnam Inc.

completely changed several times in that period. The market for office buildings is influenced by many factors including demand for space, potential return on investments from lease or sale of buildings, interest rates, government planning regulations regarding plot ratios, and fashion, factors which have nothing at all to do with the structural integrity of buildings. As Singaporean Architect Goh Chong Chia, former president of the Singapore Institute of Architects, explains, buildings in Singapore are considered replaceable, even temporary artefacts by real-estate investors, yet must be designed as if permanent.*

Whenever there is demand for new building, whether it be to replace an outdated building or an office interior, new materials are created, old materials are very often dumped in landfills and a range of environmental impacts associated with the building supply chain occur. It is worth referring again to Treloar's 1996 study[41] of the em-

*Goh Chong Chia interviewed in July 2000.

bodied energy input of retrofitting to provide an example. His study of an office building in Melbourne, Australia, showed that due to retrofitting, the total embodied energy over a projected 40-year design life accounted for 60% of the building's total energy requirements. This means that the energy required to avoid 'service life' obsolescence, in this case, of interior space, can be greater than the energy required to operate the building services over its life.

BEEs build adaptability

BEEs reduce the likelihood of their projects becoming obsolete by incorporating measures that increase a building's ability to deal with change, otherwise known as its 'adaptability'. An adaptable building is one that is built to

> '...accommodate changes in its use, its expansion or contraction of space, or major changes to its systems and envelope can be accommodated with minimal waste of resources'.[42]

The concept of building adaptability is analogous with resilience in ecosystems. Both concepts describe the ability of built and natural systems, respectively, to sustain their function in the face of change.

One key strategy BEEs use to achieve adaptability in buildings is to design components and systems in a way that makes it easy for them to be disassembled for reuse, rather than demolished when a building reaches the end of its service life. This strategy is known in the industry as *design for deconstruction.*

While many building professionals consider the adaptability of space within the building envelope, BEEs also consider adaptability at the scale of the site or surrounding environment. They ask questions such as: how rigid are the connections between the building and its surrounding environment? Can, for example, a building easily connect and disconnect from sanitation infrastructure? Is the building designed to make a lasting impression on the site? And, does it have a small enough footprint to allow the environment on the site to adapt to the building and recover when the building is removed?

Figure 7.4 shows houses designed for a mining town in a rainforest area of Irian Jaya where a modular design increases the ease with which components can be added or rearranged. The connection of the house to the site was designed to allow the natural environment to adapt to the presence of the building, while services were run into a common easement allowing easy maintenance, connection and disconnection.[43]

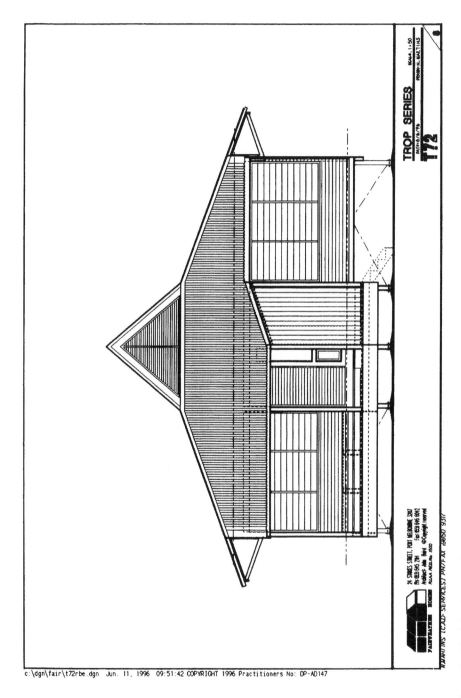

Fig. 7.4 Modular houses for Irian Jaya. This 'Tropical Series' design used prefabricated timber-framed wall and roof trusses. The simple connection details allowed for local assembly and easy shipment to the remote township in the west of Irian Jaya, Indonesia. Simple technology allows for easy construction, maintenance and eventual disassembly. Source: John Baird, Fairweather Homes Pty Ltd, Australia. (*Continued.*)

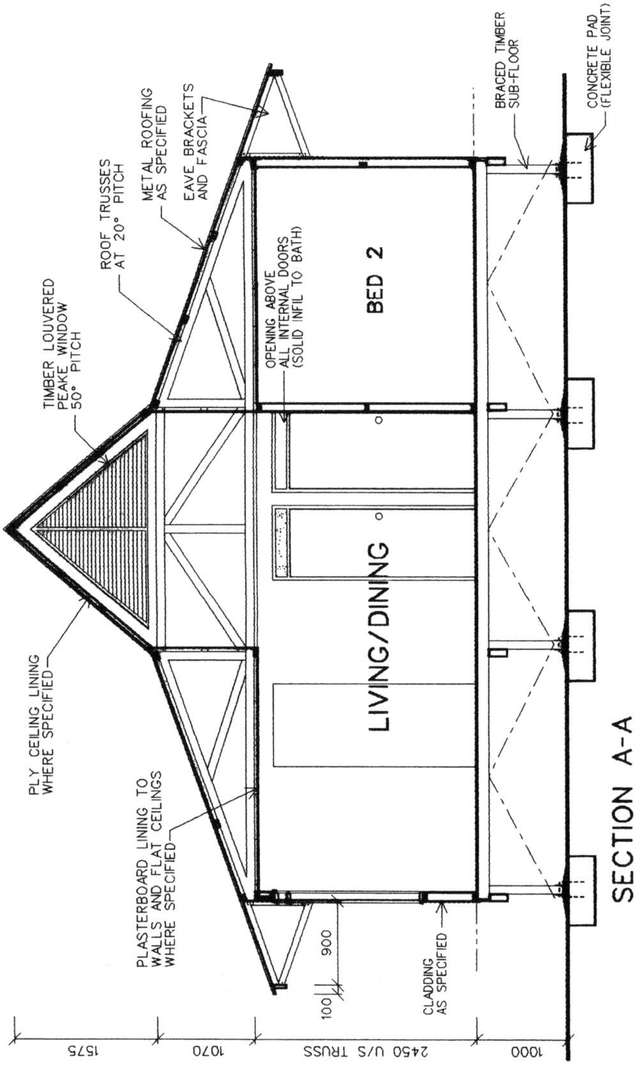

Fig. 7.4 *(Continued.)*

Labels in figure:

TIMBER LOUVERED PEAK WINDOW 50° PITCH

ROOF TRUSSES AT 20° PITCH

METAL ROOFING AS SPECIFIED

EAVE BRACKETS AND FASCIA

BRACED TIMBER SUB-FLOOR

CONCRETE PAD (FLEXIBLE JOINT)

OPENING ABOVE ALL INTERNAL DOORS (SOLID INFIL TO BATH)

BED 2

PLY CEILING LINING WHERE SPECIFIED

PLASTERBOARD LINING TO WALLS AND FLAT CEILINGS WHERE SPECIFIED

LIVING/DINING

CLADDING AS SPECIFIED

1575

1070

2450 U/S TRUSS

100   900

1000

SECTION A-A

172

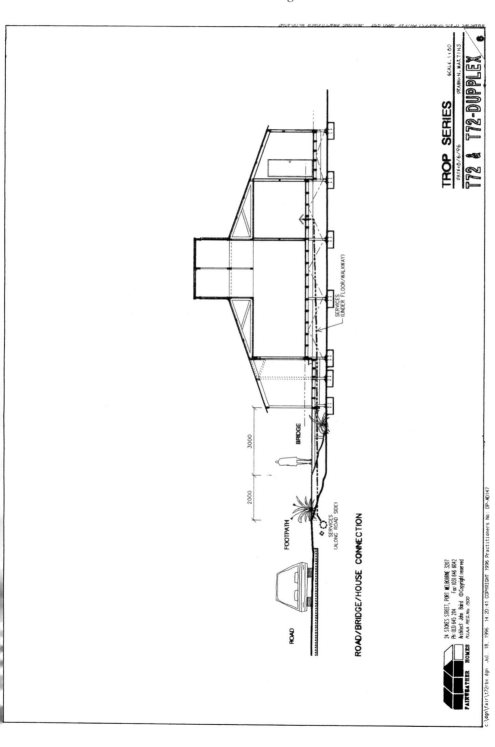

Fig. 7.4 (*Continued.*)

BEEs are cautious

Taking a precautionary approach to decision-making means adhering to the intent of the 'precautionary principle' defined as follows:

'Where there are threats of serious or irreversible environmental damage, lack of full scientific certainty should not be used as a reason for postponing measures to prevent environmental degradation'.

Decisions should therefore be guided by:

- careful evaluation to avoid, wherever practicable, serious or irreversible damage to the environment; and
- an assessment of the risk-weighted consequences.[44]

The precautionary principle is invoked when it is likely that an action will pose a serious or irreversible threat to an environment and there is uncertainty about the likely environmental consequences of a decision. It basically means that if there is any doubt about the environmental implications of a decision, then the action should not be taken until we are satisfied that environmental damage will not occur. As Deville and Harding state, applying the precautionary principle simply requires 'assessing the potential for threats of serious environmental damage'.[45]

Decisions that lead to ecology sustaining buildings as discussed previously would avoid as many negative impacts as possible and minimise unavoidable damage. However, due to the nature of most common building materials and the structure of the industry it is likely that some negative environmental impacts will occur as a result of decisions made. Because of this it is important to follow a precautionary hierarchy of decision-making, which is to:

- **Avoid** following a course of action that will lead to a serious environmental impact.
- **Reduce** Where the above is not possible, take steps to reduce the magnitude of the impact.
- **Optimise** Once impacts are reduced, take steps to optimise the benefit of the decision and continually reduce the impact over time.

Concepts of perception, feedback, learning, avoiding obsolescence, and exercising precaution are important personal and project-specific approaches to creating ecologically sustainable building. But how do BEEs contribute to positive change in environmental and social conditions beyond the building site and within industry? To answer this question we need to talk about the movements of sustainable development and sustainable construction.

Sustainable development

The term 'sustainable development' was introduced in 1980 to describe development efforts which sought to address social needs while taking care to minimise potential negative environmental impacts.[46] This concept recognised the need to maintain the natural environment in a state that still allowed humanity to fulfil its needs. The concept did, however, stress that achieving economic and social goals was dependent on, and connected with, achieving environmental goals.[47] The two words that make up the name of this concept need to be explored. While sustainable development might be easy to say, debates about what should be sustained and what constitutes appropriate forms of development are hotly contested.

Sustainability?

Sustai'n ~ ABLE *adj.* Able to keep going continuously.

The Oxford dictionary describes sustainability as the ability to 'maintain or prolong'. The question of what exactly it is that is important to maintain or prolong varies between different groups within society. Some talk of economic sustainability, some of social or political sustainability, and others of ecological sustainability. All of these concepts have been to varying degrees encompassed by the term 'sustainable development'.

Hill and Bowen[48] present a good analysis of the debate on the meaning of sustainability and how it relates to building. They point to two opposing philosophies, which the concept of sustainable development seeks to contain. One is the 'limits to growth' perspective, established by the Club of Rome in 1972, which emphasises the need to consider pollution, environmental degradation and natural resource depletion as crucial to the future of humanity.[49] This opposes the other, more traditional, 'pro-growth' philosophy, which still

175

predominates today in which sustainable development is seen as a synthesis of economic growth within the capabilities of the natural environment, leading to the view that:

> 'Anyone can sign up for sustainable development so long as it requires no specific commitment to do anything which threatens their material interests.'[50]

Sustainable development has therefore been used to synthesise opposing views. This has led some people to argue that sustainability as an ideology is wrong and is used far more by governments as a way of creating permanence in existing human systems, and would be more accurately classified as 'substitutive conservatism'.[51]

While ecological processes are fundamental, they are interdependent with economic, social and cultural systems. Humans, unlike other species, have created systems which are not considered a part of the ecosystems upon which we survive. Political systems, and cultural values, are examples of this. It is the tapestry of human society and not the web of life that cause the quest for sustainability to be complicated!

In developing countries, for example, rapid population growth and increasing urbanisation is creating environmental conditions that are a challenge to survive. Under these conditions, the physical needs of daily living become consuming and can create atmospheres of civil tension, crime and political destabilisation. Concern for the natural environment is therefore far from being a priority.

Sustainable development policy in countries such as China and Botswana, for example, emphasise economic development as a primary goal.[52,53] The rationale for this approach is that, until people have achieved a certain standard of living, they will not be in a position to consider the wide environmental implications of their development patterns. George Orwell once wrote:

> 'People with empty bellies never despair of the universe, nor even think about the universe, for that matter'.[54]

The trouble is that the styles of development that have already been established as successful in improving human health and comfort have not always nurtured the kinds of value systems, nor biophysical side-effects, which provide for ecological sustainability.[55]

Whether we subscribe to the pro-growth economic perspective, a social equity perspective, both or neither, the ability to even hold a view relies on ecological sustainability. No matter what else is missing, natural systems must continue to provide the basic life-support systems in which human activities can take place. Without a functioning biosphere social, cultural and economic sustainability would be impossible. It was based on this observation that the concept of ecologically sustainable development (ESD) was developed.

A well-known definition of sustainable development was published in 1987 in the World Commission on Environment and Development (WCED) report *Our Common Future*. It states that: 'Sustainable development is development which meets the needs of the present without compromising the ability of future generations to meet their own needs.'[56] The first thing to notice about this definition is that it talks about needs. Considering that access to natural resources and the free products of functioning ecosystems have been essential for progress thus far, it follows they be sustained in order to ensure that the needs of present and future generations are met. However, needs vary with changing situations. The purpose of adding the word 'ecologically' to sustainable development is to emphasise that our prosperity is interdependent with that of natural systems.

A recent report into the global status of sustainable construction activity found differences in the perception of what is required for ecologically sustainable development, between developed and developing countries.[57] The report found that developed countries perceive sustainability issues as those that threaten existing living conditions and quality of life. These issues include pollution, over-consumption and reduced access to natural resources, inequitable distribution of wealth, and welfare. Developing countries that participated in the research perceived sustainability issues as those issues that impede efforts to improve the living conditions and quality of life for their people. Providing water and sanitation, housing and schools are major priorities. Common to both is the recognition of the need for changes in current patterns of social and economic activity.

Both developed and developing countries also experience common impediments to sustainable development and construction initiatives. These are generally barriers to instigating change and include complicated or unintegrated environmental protection and development planning bureaucracies, the competing values of financial and environmental wealth, and people's attitudes to change and patterns of consumption. There is also shared concern for global environmental loadings such as greenhouse gas emissions, environmental pollution and natural resource depletion. All countries involved in the study were concerned with the relatively slow pace with which innovations contributing to sustainable development and construction are being adopted.

The concept of ESD does not describe some future static state but a way of ensuring that human progress continues. The only thing that is in fact constant about the world is that nothing ever remains the same. Change is constant. Many researchers, activists and policy-makers have called for changes in patterns of resource consumption to achieve sustainable development.[58,59,60] In addition, there have been calls for a change in the way we relate with and value nature and a reintegration of spiritual and scientific worldviews.[61]

While there is little disagreement about the need to sustain functioning natural systems, there is widespread debate about the form of 'development' required to achieve this aim. As Tisdall (1993) states, the concept of sustainability is considered as a dynamic integration of economics, sociology and ecology:

> 'Sustainable development may require sustainable patterns of economic exchange as well as sustainability of political and social structures and sustainability of community...Thus the concept of sustainability now rises in many contexts involving development.'[62]

Tisdall points out that sustainable development is a concept made up of the ideas of sustainability and development and that these ideas are not mutually inclusive.

Development?

The dictionary definition of development is 'growth', 'evolution' and a 'stage of advancement'.[63] Development has also been defined as 'modification to the biosphere to satisfy human needs'.[64] This human-centred view is implied in the previous definition of sustainable development by the World Commission on Environment and Development.[65]

With more specific reference to building, development has also been described as an 'outward expansion of undeveloped land'.[66] This second definition certainly implies growth but does not necessarily equate with evolution or advancement. Development as evolution suggests an unfolding natural process of improvement where development occurs via learning from past patterns and adapting to new conditions. Development as creation, on the other hand, implies consciously making some kind of product our outcome.

Growth in economic activity has also been termed 'development'. Economic activity, which involves the production and consumption of products, may or may not lead to gradual improvement in environmental conditions and, therefore, may or may not be evolutionary. It has been argued that certain forms of economic development are, in fact, counter evolutionary.[67] Economist Herman Daly goes further and argues that economic growth is no longer possible because our current economic system assumes no limits in nature, and that what we take from nature can be made useful in natural systems again when it is no longer useful to humans.[68]

Taking materials from nature and turning them into buildings, for example, has been thought of as substituting 'natural capital' for 'man-made' capital and as an ecologically neutral activity.[69] This perspective is based on the questionable assumption that, once removed from nature, materials from constructed environments remain available for use by, or provide services for future generations. This view of using 'natural capital' ignores the ecological damage caused by natural resource consumption on a massive scale. This is a result of the false assumption that, since natural capital can be 'liquidated' for use by humans, so 'human capital' can be liquidated again for use by nature.[70]

BEEs think of development more in the evolutionary sense and therefore see development as a process of changing environmental conditions for the better. Ecologically sustainable building therefore becomes:

> A process of building that can be sustained by natural systems, and which in turn helps sustain natural systems.

Note that this definition does not stipulate the type of action that might lead to positive change and therefore may or may not involve constructing something.

As Part I of this book showed, the way we have gone about building can neither be sustained by natural systems, nor has it helped sustain them. We have changed the natural environment in order to create the built environment without always considering the possible effects in nature. The accumulation of our impact is now causing natural systems to change in ways that threaten the viability of the methods we currently have of improving human conditions.

Part of the process of ecologically sustainable building is therefore changing ecologically unsustainable current practices, and adopting ecologically sustainable ones. BEEs call this movement of change within the building industry *sustainable construction.*

Sustainable construction

Sustainable construction is not a particular future state. Rather, it is a process by which building development can progress from a perceived unsustainable present, to a more sustainable future. As noted by Gardner[71] 'we are often not starting with the conditions that we would want to sustain'. Principles of sustainable construction therefore, help describe a journey, not a destination.

This journey is a process of changing the building industry so that it becomes ecologically sustainable. Unfortunately, the building industry is traditionally slow to change practices. Many building firms adopt a defensive attitude to calls for change in practice. Sustainable construction is therefore seen as a process for driving industry innovation toward ecological sustainability. The various stages of this campaign are shown in Fig. 7.5.

The International Council for Research and Innovation in Building and Construction (CIB) sees this as a process of helping the industry to reduce resource consumption and resulting environmental loading. It advocates achieving these aims through:

• promoting resource efficiency;
• reducing use of high-quality drinking water;
• selecting materials on environmental performance;
• contributing to sustainable urban environments.[72]

All of these issues and responses present a developed-world perspective of problems affecting the sustainability of construction. Not all journeys towards a more sustainable built environment begin with the same set of conditions and this needs to be recognised in developing a sustainable construction approach. Creating more sustainable housing in the slums of Mexico City, for example, would be a vastly different process to increasing the sustainability of housing in Beverley Hills.

BEEs know that the success of sustainable construction initiatives is directly related to how well contextual issues are considered.[73] The context in which a building development takes place will, therefore, have a fundamental influence on whether it is or is not sustainable. In addressing this issue the CIB sees sustainable construction as an important element in creating ecologically sustainable urban environments that respond to local ecological, social and economic conditions.

In order to progress sustainable construction the CIB calls for the development of 'external drivers' including regulations, increasing energy prices, more information, education and tools, tax incentives,

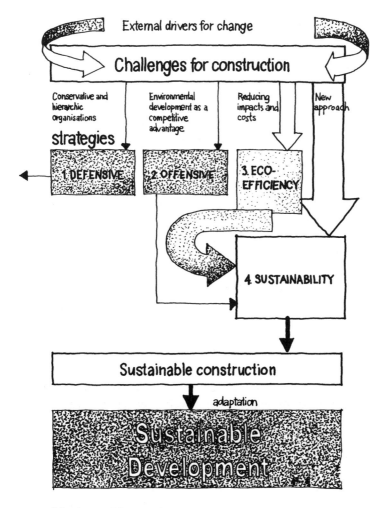

External drivers for change

Challenges for construction

Conservative and hierarchic organisations

Environmental development as a competitive advantage

Reducing impacts and costs

New approach

strategies

1. DEFENSIVE

2. OFFENSIVE

3. ECO-EFFICIENCY

4. SUSTAINABILITY

Sustainable construction

adaptation

Sustainable Development

Various Strategies

Fig. 7.5 Stages in the sustainable construction campaign. The direction of innovation in the construction industry globally is toward sustainability. The challenge is for practitioners and companies to move from (1) defensive positions of 'We'll change if you regulate', through (2) offensive positions that recognise the competitive advantages of environmental development, perhaps offering environmental services but without changing core business practice or mind-set, to (3) eco-efficiency, where they don't just offer environmental services, but look at ways of changing their business practices so they are less damaging to the environment. Beyond eco-efficiency is (4) sustainability, where sources of pollution and waste are eliminated and new ways of working with natural systems are adopted. Social and economic reforms are considered and the values of equity and co-operation, rather than growth and competition, are included in decision-making. Source: International Council for Research and Innovation in Building and Construction (CIB) (1999) *Agenda 21 on sustainable construction.* CIB Report Publication 237, July. Rotterdam. p.90.

and demonstration projects. While these measures are necessary, BEEs don't wait around for someone else to act as an external driver. BEEs take the wheel because they know how to drive sustainable construction at the building project level.

Hill and Bowen[74] provide a model for implementing concepts of sustainable construction to a building project. Their 'four pillars' model is shown in Fig. 7.6 and describes project goals that if adhered to might assist the building in contributing to sustainable construction. The four pillars consist of social, economic, biophysical and technical sustainability goals.

BEEs know that choosing sustainable construction goals for a project requires an investigation of the issues arising from the context in which the building is taking place. They also know that they need to involve people who are likely to be affected by the project in order to understand local concerns. Because BEEs are life-cycle thinkers they know that principles need to be chosen to influence the ecological sustainability of a project over time. BEEs therefore follow a six-part process when choosing sustainable construction goals from the 'pillars'. These six steps are:

(1) establish existing ecological, social and economic conditions;
(2) determine the possible effects of development alternatives;
(3) decide which impacts are most significant;
(4) choose project goals in consultation with the people affected;
(5) assess/monitor adherence to project goals; and
(6) provide feedback.[75]

Establish existing ecological, social and economic conditions

The sustainability of a building project will be determined ultimately by how well it responds to issues arising from the ecological, social, cultural and economic variables that describe its context. Determining the potential impacts of different building development options requires BEEs to make decisions about the likely combined effects of issues that have traditionally been treated separately. This process will involve BEEs in organising and/or conducting community consultation, multi-disciplinary and multi-stakeholder input into project planning, and applying life-cycle and systems thinking.

BEEs try to determine the sensitivity of the existing environment to change. This might involve conducting habitat, demographic, cultural and archaeological surveys of a building site or affected region, considering employment rates, neighbourhood character and the marginal capacity of local infrastructure. BEEs would also identify transportation flows and choices, determine ambient noise levels and

```
                          PROCESS-ORIENTED PRINCIPLES OF SUSTAINABLE
                                         CONSTRUCTION
```

Over-arching principles indicating approaches to be followed in evaluating the applicability and importance of each 'pillar', and its associated principles, to a particular project.

○ Undertake prior assessments of proposed activities	○ Recognize the necessity of comparing alternative courses of action	○ Establish a voluntary commitment to continual improvement of performance
○ Timeously involve people potentially affected by proposed activities in the decision-making process	○ Utilize a life cycle framework	○ Manage activities through the setting of targets, monitoring, evaluation, feedback and self-regulation of progress
○ Promote interdisciplinary collaborations and multi-stakeholder partnerships	○ Utilize a systems approach	
	○ Exercise prudence	
	○ Comply with relevant legislation and regulations	○ Identify synergies between the environment and development

PILLAR ONE: SOCIAL SUSTAINABILITY

○ Improve the quality of human life, including poverty alleviation

○ Make provision for social self determination and cultural diversity in development planning

○ Protect and promote human health through a healthy and safe working environment

○ Implement skills training and capacity enhancement of disadvantaged people

○ Seek fair or equitable distribution of the social costs of construction

○ Seek equitable distribution of the social benefits of construction

○ Seek intergenerational equity

PILLAR TWO: ECONOMIC SUSTAINABILITY

○ Ensure financial affordablity for intended beneficiaries

○ Promote employment creation and, in some situations, labour intensive construction

○ Use full-cost accounting and real-cost pricing to set prices and tariffs

○ Enhance competitiveness in the market place by adopting policies and practices that advance sustainability

○ Choose environmentally responsible suppliers and contractors

○ Invest some of the proceeds from the use of non-renewable resources in social and human-made capital, to maintain the capacity to meet the needs of future generations

PILLAR THREE: BIOPHYSICAL SUSTAINABILITY

○ Extract fossil fuels and minerals, and produce persistent substances foreign to nature, at rates which are not faster than their slow redeposit into the Earth's crust

○ Reduce the use of the four generic resources used in construction, namely, energy, water, materials and land

○ Maximize resource reuse, and/or recycling

○ Use renewable resources in preference to non-renewable resources

○ Minimize air, land and water pollution, at global and local levels

○ Create a healthy, non-toxic environment

○ Maintain and restore the Earth's vitality and ecological diversity

○ Minimize damage to sensitive landscapes, including scenic, cultural, historical, and architectural

PILLAR FOUR: TECHNICAL SUSTAINABILITY

○ Construct durable, reliable, and functional structures

○ Pursue quality in creating the built environment

○ Use serviceability to promote sustainable construction

○ Humanize larger buildings

○ Infill and revitalize existing urban infrastructure with a focus on rebuilding mixed-use pedestrian neighbourhoods

Fig. 7.6 Four pillars of sustainable construction. Sustainable construction is a dynamic integration of social sustainability, biophysical sustainability, economic sustainability and technical sustainability. This model recognises that human society and the biosphere are interdependent and that changes in either affect both. Sustainable construction advocates using construction as a catalyst for positive change in all four 'pillars'. The 'process oriented principles' are ways of implementing the conditions described in each 'pillar' during the building development process. Source: Hill, R.C. & Bowen, P.A. (1997) Sustainable construction: principles and a framework for attainment. *Construction Management and Economics* **15** 228. Fig. 1.

Some contextual issues for ecologically sustainable building

- Ecological, cultural, historical and recreational value of development land
- Ambient atmospheric characteristics
- Transportation, vehicular access, cycle paths, public transport
- Scarcity of development land
- Availability of existing buildings with potential for renovation
- Adequacy of water supply
- Adequacy of municipal services infrastructure to meet marginal demand
- Solar availability
- Emissions from regional generation of electricity
- Cultural and historic value of local built environment

Source: Based on 'Context' criteria in the Green Building Tool version 2K 1.75, http://www.greenbuilding.ca/iisbe/gbc2k2/gbc2k2-start.htm

local air and water quality. The box above presents some of the contextual issues that BEEs consider.

Determine the possible effects of development alternatives and which impacts are most significant

At the scale of determining the environmental effects of entire developments, a number of assessment techniques can be used. A development-wide evaluation technique commonly applied to major projects is environmental impact assessment (EIA).

In many countries environmental protection legislation requires an environmental impact assessment to be conducted on major projects to identify likely combined environmental effects. The issues that determine whether a project should be subject to an environmental impact assessment as agreed by the Australian and New Zealand Environment and Conservation Council (ANZECC) in 1993[76] are:

- the character of the receiving environment;
- the potential impacts of the proposal;
- the resilience of the environment to cope with change;
- the confidence in prediction of impacts;
- the degree of public interest;

- presence of planning or policy framework or other procedures which provide mechanisms for managing potential environmental impacts;
- other statutory decision-making processes which may provide a forum to address the relevant issues of concern.

Environmental impact assessment is a social process developed to examine the potential environmental impacts of a project. Its main function is to determine environmental risks posed by a development alternative and to decide whether to proceed with the development and under what conditions.[77]

Conducting an environmental impact assessment requires a scoping phase and then the preparation of an environmental impact statement (EIS) and writing an assessment report. The scoping phase determines the issues that will be investigated during the preparation of the EIS. A consultative committee made up of community and environmental groups, the representatives of relevant government departments, local government representatives, and the proponent of the development guides the scoping phase.

The preparation of an EIS is carried out in four phases. These are:

(1) describing the development and the environment;
(2) predicting the impacts (magnitude and significance);
(3) evaluating the impacts;
(4) reviewing the EIS and preparing the assessment report.

This work is normally conducted by a multidisciplinary team of consultants. Both the consultative committee and the local community are consulted during the EIS and given a number of opportunities to review preliminary findings and make submissions.

An environmental impact assessment process provides a consultative approach to identifying and determining the combined effects of sustainability issues. Its use is, however, generally restricted to major development projects.

Choose project goals in consultation with the people affected

Once issues have been identified and combined effects determined, a decision is made about whether to accept or reject a development alternative and, if accepted, what level of environmental performance is required to ensure that impacts are mitigated. Principles of sustainable construction can be useful directions for describing in broad terms how a building development should affect the environment.

Sustainable construction principles can also become the framework for the building design brief.

Assess or monitor adherence to project goals and provide feedback

Once sustainable construction principles are selected and incorporated into the building design brief, they become important criteria for environmental performance assessment. Various tools and methods can be used to assess the potential performance of the building design for compliance with sustainable construction principles.

Design briefs that include performance requirements for ecologically sustainable development have been applied to landmark projects such as the Sydney Olympic Village.* On this project, design teams had to demonstrate how their design complied with the performance requirements, and used methods such as life-cycle assessment and energy modelling to improve their designs.

Environmental assessment tools will not address all of the aims of sustainable construction. However, they help BEEs contribute to ecologically sustainable building by:

- including environmental performance assessment in the design process, providing an opportunity to improve the effectiveness of the design team by including a wider range of expertise in a systematic way;
- introducing environmental assessment tools into a design process, providing a process by which designers can learn about the environmental impact of buildings and the integrated issues of sustainability;[78] and
- introducing systematic assessment of design options, providing a documented process demonstrating compliance to tender conditions or to client environmental performance goals.

Environmental assessment of building design varies in scope and application depending on the stage of the design at which it is applied, the amount of time required for assessments to be carried out, the existing knowledge of the design team, access to information, and available financial resources. In most cases environmental assessment is intended to help support design decisions. It therefore follows that for this to take place, the design decision-making process itself must be structured in a way that enables the inclusion of environmental information.

*This information is contained in the ESD design guidelines for the Sydney Olympic Village project, provided by the project developers Mirvac/Lend Lease.

In summarising the debate about sustainability and the direction of socio-economic and ecological change influenced by ecologically sustainable building, it is important to remember that sustainability is a process, not an outcome. The nature of both human and natural systems is that they change over time. Adapting to this change is therefore necessary. Sustainable development and the construction industry's response – sustainable construction – require rethinking current practice. They don't answer our questions about what the future should be, but they provide directions for changing the aspects of current lifestyles and building practices that undermine our ability to live in harmony with each other and nature.

Meeting the requirements of sustainable development and sustainable construction may not be possible using existing management and administrative systems. It may not even be possible within existing value systems. What is important is for us to continue to debate how the future should be and to change current practices so that we adhere to the fundamental laws of nature, and meet the challenges of change. The future might be able to take care of itself if we take care to support the ecological health of the planet and the well-being of our communities.

We have now investigated the basic knowledge that BEEs apply to this work. The next chapter summarises this knowledge to describe what we BEEs now know about ecologically sustainable building.

Remember: The only thing that's certain is that everything is changing.

References

1 King, M.L. Jnr, Rev. (1968) *The Trumpet of Conscience*. Harper & Row, New York. As quoted by Flanagan, M. (2001) "A Voice of Our Time" perspective. *The Age Newspaper* 26 November 2001. p.13. Melbourne, Australia.

2 Brown, L. (1995) The Acceleration of History. *Vital Signs 1995–1996*, World Watch Institute, Earthscan Publications Ltd., London, UK.

3 World Resources Institute (1996) *World Resources 1996–1997: a guide to the global environment*. Oxford University Press, New York.

4 UNEP (2001) Summary for Policy Makers: A report of Working Group One of the Intergovernmental Panel on Climate Change. *Climate Change Synthesis Report* United Nations, Geneva, Switzerland. pp.2–4.

5 World Watch Institute (2000) Earth Day 2000: A 30-year report. *World Watch* March/April. World Watch Institute Washington DC, USA. p.11.

6 Commonwealth Science and Industry Research Organisation (1998) *Waste Water Treatment & Management in Australia*. CSIRO Publishing, Melbourne, Australia.

7 *Concise Oxford Dictionary* (1976) Sixth Edition Second Imprint. Sykes, J.B. (ed.) Oxford University Press, London, UK.

8 Capra, F. (1997) *The Web of Life: a new synthesis of mind and matter.* Flamingo, Harper Collins, London. p.102.

9 Bateson, G. (1979) *Mind and Nature: a necessary unity.* E.P. Dutton, New York.

10 Lister, N. & Kay, J. (1999) Celebrating Diversity: adaptive planning and biodiversity conservation. In: Bocking S. (ed.) *Biodiversity in Canada: an introduction to environmental studies.* Broadview Press, Peterborough, Canada. pp.189–218.

11 Capra, F. (1997) *The Web of Life: a new synthesis of mind and matter.* Flamingo, Harper Collins, London. [and other references]

12 Fuller, R.B. (1981) *Critical Path.* St. Martin's Press, New York. p.150.

13 Senge, P.M. (1992) *The Fifth Discipline: the art and practice of the learning organization.* Random House Australia, Sydney.

14 Kong, L., Yuen, B., Sodhi, N. & Briffett, C. (1999) The construction and experience of nature: perspectives of urban youths. *Tijdschrift voor Economische en Sociale Geografie* **90** (1) 3–16.

15 Holling, C. (1986) The resilience of terrestrial ecosystem: local surprise and global change. In: Clark, W. & Munn, R. (eds) *Sustainable Development of the Biosphere.* Cambridge University Press, Cambridge, UK.

16 Peterson, G., Allen, C. & Holling, C. (1998) Ecological Resilience, Biodiversity, and Scale. *Ecosystems* Vol 1. New York, Inc. USA. pp.6–18.

17 Smith, S. (1968) Species succession and fishery exploitation in the Great Lakes. *Journal of the Fisheries Research Board of Canada* **25** 667–693.

18 Holling, C. (1973) Resilience and Stability of Ecologically Systems. Research monograph, Institute of Resource Ecology, University of British Columbia, Vancouver, Canada.

19 United Nations Development Program (UNDP), United Nations Environment Program (UNEP), World Bank & World Resources Institute (2000) *A Guide to World Resources 2000–2001: people and ecosystems: the fraying web of life.* World Resources Institute, Washington DC. p.7.

20 Cole, R.J. & Lafreniere (1997) Environmental information frameworks: linking research with building design. *Buildings and the Environment,* Proceedings Second International Conference CIB Task Group 8. 9–12 June, Paris, France. Vol 2 pp.33–40.

21 Walters, C. (1986) *Adaptive Management of Renewable Resources.* Macmillan Publishing, New York. pp.1–12.

22 Odum, H.T. (1996) *Environmental accounting: EMERGY and environmental decision making.* John Wiley & Sons, New York.

23 Lister, N. & Kay, J. (1999) Celebrating Diversity: adaptive planning and biodiversity conservation. In: Bocking S. (ed.) *Biodiversity in Canada: an introduction to environmental studies.* Broadview Press, Peterborough, Canada. pp.189–218.

24 Senge, P.M. (1992) *The Fifth Discipline: the art and practice of the learning organization.* Random House, Australia, Sydney.

25 Gelder, J. (1998) Using School Buildings to Teach Environmentally Sustainable Design. In: *BDP Environment Design Guide* General Issues Paper 23, November. Melbourne, Australia.

26 Lingard, H., Graham, P. & Smithers, G. (2000) Employee perceptions of the solid waste management system in a large Australian contracting organisation: implications for company policy implementation. *Construction Management and Economics* **18** 383–393.

27 Lingard, H., Smithers, G. & Graham, P. (2001) "Improving solid waste reduction and recycling performance using goal setting and feedback" *Construction Management & Economics* **19** 809–817.

28 Lingard, H., Graham, P. & Smithers, G. (2000) Employee perceptions of the solid waste management system in a large Australian contracting organisation: implications for company policy implementation. *Construction Management and Economics* **18** 383–393.

29 Kay, J., Regier, H., Boyle, M. & Francis, G. (1999) An Ecosystems Approach for Sustainability: addressing the challenge of complexity. *Futures* **31**(7) 721–742.

30 Smith, M., Whitelegg, J. & Williams, N. (1999) *Greening the Built Environment*. Earthscan Publications, London.

31 Pimm, S. (1984) The Complexity and Stability of Ecosystems. *Nature* **307** 321–326.

32 Kibert, C.J. (2000) Construction ecology and metabolism. In: Boonstra, C., Rovers, R. & Pauwels, S. (eds) *International conference sustainable building 2000 proceedings*. 22–25 October, Maastricht, Netherlands, Aeneas Technical Publishers. pp.177–179.

33 Peterson, G., Allen, C. & Holling, C. (1998) Ecological Resilience, Biodiversity, and Scale. *Ecosystems* Vol 1. New York, Inc. USA.

34 Holling, C. (1973) Resilience and Stability of Ecologically Systems. Research monograph. Institute of Resource Ecology, University of British Columbia, Vancouver, Canada.

35 Schindler, D. (1990) Experimental perturbations of whole lakes as tests of hypotheses concerning ecosystem structure and function. *Oikos* **57** 25–41.

36 Golton, B. (1994) Affluence and the ecological footprint of dwelling in time – a Cyprus perspective. *First world conference on sustainable construction*, 7–9 November, CIB TG16, Tampa, Florida. Cited in: Smith, M., Whitelegg, J. & Williams, N. (1999) *Greening the built environment*. Earthscan Publications, London. p.73.

37 Duffy, F. (1990) Measuring Building Performance. *Facilities* May. Cited in: Roodman, D.M. & Lenssen, N. (1995) *A Building Revolution: how ecology and health concerns are transforming construction. World Watch Paper No. 124*. World Watch Institute, Washington DC, USA. p.17.

38 Brand, S. (1994) *How Buildings Learn: what happens after they're built*. Penguin Books, New York.

39 Kibert, C.J. (2000) Deconstruction as an essential component of sustainable construction. In Boonstra, C., Rovers, R. & Pauwels, S. (eds) *International conference sustainable building 2000 proceedings*. 22–25 October, Maastricht, Netherlands, Aeneas Technical Publishers. p.89.

40 Smith, M., Whitelegg, J. & Williams, N. (1999) *Greening the Built Environment*. Earthscan Publications, London.

41 Treloar, G.J. (1996) *The Environmental Impact of Construction: a case study*. Australia and New Zealand Architectural Science Association, Sydney.

42 Grammnos, F. & Russell, P. (1997) Building adaptability: a view from the future. In: *Buildings and the environment, second international conference; environmental management, environmental strategies,* 9–12 June, Paris, **2,** Centre Scientifique et Technique du Butimat, 19-26.

43 Fairweather Homes (1996) *500 House Development for P.T. Freeport, Indonesia at Kuala Kencana.* Fairweather Homes Design Proposal, Fairweather Homes, Phillip Island, Australia.

44 Intergovernmental Agreement on the Environment (1992) Heads of Government in Australia, May as cited in: Deville, A. & Harding, R. (1997) *Applying the Precautionary Principle.* Federation Press, Leichhardt, Australia. p.13.

45 Deville, A. & Harding, R. (1997) *Applying the Precautionary Principle.* Federation Press, Leichhardt, Australia. p.25.

46 International Union for the Conservation of Nature and Natural Resources (IUCN) (1980) *World Conservation Strategy.* IUCN, Gland, Switzerland.

47 Hill, R.C., & Bowen, P.A. (1997) Sustainable construction: principles and a framework for attainment. *Construction Management and Economics* **15** 223–239.

48 Hill, R.C., & Bowen, P.A. (1997) Sustainable construction: principles and a framework for attainment. *Construction Management and Economics* **15** 223–239.

49 Stockdale, J. (1989) Pro-growth, limits to growth, and a sustainable development synthesis. *Society and Natural Resources* **2**(3) 163–176. Cited in Hill, R.C., & Bowen, P.A. (1997).

50 Chen, J., Chambers, D., Wills, D. & Kaixun Sha (1998) Sustainable Development and Sustainable Construction in China. CIB Task Group 16. *Compendium on Sustainable Construction.* CIB Triennial Conference, Gavle, Sweden. October.

51 Treanor, P. (1997) *Why Sustainability is Wrong.* December. http://web.inter.nl.net/users/Paul.Treanor/sustainability.html.

52 Chen, J., Chambers, D., Wills, D. & Kaixun Sha (1998) Sustainable Development and Sustainable Construction in China. CIB Task Group 16. *Compendium on Sustainable Construction.* CIB Triennial Conference, Gavle, Sweden. October.

53 Ngowi, A.B. & Hunt, G. (1998) Sustainable Development and Construction in Botswana. CIB Task Group 16. *Compendium on Sustainable Construction.* CIB Triennial Conference, Gavle, Sweden. October.

54 Orwell, G. (1962) *Inside the Whale and other essays.* Penguin Books Ltd. The Whitefriars Press Ltd, London, UK.

55 Knudtson, P. & Suzuki, D. (1992) *Wisdom of the Elders.* Allen & Unwin, Sydney.

56 World Commission on Environment and Development (WCED) (1987) *Our Common Future.* Oxford University Press. UK.

57 International Council for Research and Innovation in Building and Construction (CIB) (1998) Compendium of Sustainable Development & Construction. CIB Task Group 16. *Compendium on Sustainable Construction.* CIB Triennial Conference, Gavle, Sweden. October.

58 Meadows, D., Meadows, D., Randers, J. & Behrens, W. (1972) *The Limits to Growth: A Report for the Club of Rome's Project on the Predicament of Mankind.* Universe Books, New York.

59 Meadows, D., Meadows, D. & Randers, J. (1992) *Beyond the Limits: confronting global collapse, envisioning a sustainable future.* Chelsea Green Publishing Co. White River Junction, Vermont, USA.

60 Weizsäcker, E.von, Lovins, A.B. & Lovins, L.H. (1997) *Factor Four: doubling wealth, halving resource use: the new report of the Club of Rome.* Allen & Unwin, Sydney.

61 Knudtson, P. & Suzuki, D. (1992) *Wisdom of the Elders.* Allen & Unwin, Sydney.

62 Tisdall, C. (1993). *Environmental Economics.* University Press, Cambridge, UK.

63 *Concise Oxford Dictionary* (1976) Sixth Edition Second Imprint. Sykes, J.B. (ed.) Oxford University Press, London, UK.

64 International Union for the Conservation of Nature and Natural Resources (IUCN) (1980) *World Conservation Strategy.* IUCN, Gland, Switzerland.

65 World Commission on Environment and Development (WCED) (1987) *Our Common Future.* Oxford University Press. UK.

66 Harvey, J. (1981) *The Economics of Real Property.* Macmillan Publishers Ltd, Hampshire, UK.

67 Schumacher, E.F. (1973) *Small Is Beautiful: a study of economics as if people mattered.* Vintage Books, London, UK. Reprint 1993.

68 Daly, H. (1996) *Beyond Growth.* Beacon Press, Boston.

69 Solow, R. (1993) An Almost Practical Step to Sustainability. Reviewed in *Resources* Vol. 110 Resources for the Future Washington DC, USA. Cited in Hill, R.C., & Bowen, P.A. (1997) Sustainable construction: principles and a framework for attainment. *Construction Management and Economics* **15** 225.

70 Solow, R. (1993) An Almost Practical Step to Sustainability. Reviewed in *Resources* Vol. 110 Resources for the Future Washington DC, USA. Cited in Hill, R.C., & Bowen, P.A. (1997) Sustainable construction: principles and a framework for attainment. *Construction Management and Economics* **15** 225. p.5

71 Gardner, J.E. (1989) Decision making for sustainable development: selected approaches to environmental assessmentand management. *Environmental Impact Assessment Review,* Canada. pp.337–366.

72 International Council for Research and Innovation in Building and Construction (CIB) (1999) *Agenda 21 on sustainable construction.* CIB Report Publication 237, July. Rotterdam.

73 Dubose, J.R. (1994) *Sustainability as an Inherently Contextual Concept: some lessons from agricultural development.* MSc thesis. Georgia Institute of Technology, June. Atlanta, USA.

74 Hill, R.C., & Bowen, P.A. (1997) Sustainable construction: principles and a framework for attainment. *Construction Management and Economics* **15** 223–239.

75 Graham, P. (1997) *Methods for Assessing the Sustainability of Construction and Development Activity* – Dissertation for Master of Applied Science (Building) RMIT University, Melbourne Australia, August.

76 Department of Planning & Development (1995) Guidelines for Environmental Impact Assessment and the Environment Effects Act, Melbourne, Australia, April.

77 Thomas, I. (1998) *Environmental impact assessment in Australia: theory and practice.* The Federation Press, NSW, Australia.

78 Malin, N. 1998 The Green Building Adviser: design team on a disc. *Green Building Challenge '98* CDROM Proceedings: GBC'98 International Conference on the Performance Assessment of Buildings. October. Vancouver.

8 SUMMARY: WHAT DO BEES KNOW NOW?

Thermodynamics

In accordance with the first law of thermodynamics, BEEs first avoid unnecessary consumption of materials and energy, and second, reduce consumption by maximising efficient use. Consumption in this sense refers both to building materials themselves and all waste associated with their life cycle. In order to accord with the second law of thermodynamics, environmentally aware building professionals first reuse, then refurbish and recycle materials, components and buildings. A feature of ecosystems is that they are structured to turn disorganised energy from the sun into organised chemical energy available for use. In this way ecosystems also generate resources or 'food' for the system through autocatalytic feedback, as explained by the fourth law of thermodynamics. Ecologically sustainable buildings emulate this feature by generating resources. The laws they apply to decision-making are therefore to:

- consume resources no faster than the rate at which nature can replenish them (first law of thermodynamics);
- create systems that consume maximum energy-quality (second law of thermodynamics); and
- create and use byproducts that are nutrients or raw materials for resource production (fourth law of thermodynamics).

BEEs in this way create buildings that avoid waste and create resources at all phases of their life cycle, incorporate secondhand and recycled materials, and thus help create 'closed loop' industries.

193

Change

At the beginning of the chapter we asked some important questions about how BEEs think about change in order to contribute to ecological sustainability. They were:

- How do BEEs determine what kind of change they need to initiate?
- In a dynamic world is the best way to contribute to ecological sustainability to make buildings that resist change or buildings that absorb change?
- What kind of change should be initiated in order to change current building practices so that they contribute to ecological sustainability?

How do BEEs determine what kind of change they need to initiate?

BEEs' choice of action depends on how accurately they perceive the change taking place in their environment and the cause. BEEs first observe what is happening to the whole system, knowing that ecological change is not a linear process of cause and effect. They look for environmental issues occurring at local, regional, continental and global scales. Judgements about the relative importance of issues are then made, based on an assessment of the development context, and in collaboration with people likely to be affected by a building project. BEEs apply a life-cycle perspective in order to understand the environmental impacts of decisions over time.

The ecological sustainability of their project is the filter through which BEEs make decisions. They therefore choose courses of action that reduce ecological impacts and restore ecological functions. BEEs know that ecosystems must be resilient to continue to function despite change and that the greater their biodiversity, the greater their resilience. BEEs know that even the most resilient ecosystems have a threshold or carrying capacity that can be breached if ecological impacts accumulate. BEEs therefore ensure that their projects protect ecosystems by minimising the number and magnitude of its associated ecological impacts. If they are unsure of the likely ecological effects of a decision, they apply the precautionary principle and postpone action until they can be sure that the outcome will be positive.

In a dynamic world is the best way to contribute to ecological sustainability to make buildings that resist change or buildings that absorb change?

Most buildings require structural stability and functional resilience. BEEs understand that different parts of a building experience different rates of change. Rather than making buildings in which all components resist change, they make buildings that allow for easy access to faster cycling materials. BEEs also apply their life-cycle thinking to allow for deconstruction and reuse of building elements. The guiding principle for BEEs is to create buildings that can adapt non-destructively to change.

What kind of change should be initiated in order to change current building practices so that they contribute to ecological sustainability?

BEEs know that the construction industry is one of the world's most resource-hungry and polluting entities. BEEs participate at a project level in sustainable construction, by implementing project goals that describe social, biophysical, economic and technical modes of project delivery. They set goals to enhance a project's contribution to ecological sustainability and its applicability to the social and economic context in which it is taking place. BEEs use these performance goals as the basis for setting environmental performance targets which they can monitor and which form the basis for feedback to workers and other project stakeholders. By adopting this approach BEEs help drive industry attitudes to environmental innovation from the defensive toward the sustainable.

What is next?

In this book we have taken a journey from understanding how and why building affects the environment by finding out what BEEs know about interdependency. We then looked at the fundamental conditions for ecological sustainability described by laws of thermodynamics and now know how change influences BEEs' decision-making. We have, if you like, collected all of our cells of honeycomb and built our hive. All that remains is for us to fly outside our structure and see what the finished product looks like. The next chapter integrates the knowledge and understanding we have gained so far so that we can see the whole picture.

More information

Thermodynamics

Energy Basis for Man and Nature (1976) Odum, H.T. & Odum, E. Mc-Graw Hill, New York.

Material circulation, energy hierarchy, and building construction (2002) Odum, H. Chapter 2 in Kibert, J., Sendzimir, J., Guy, B. (eds) (2002) *Construction Ecology Nature as the basis for green buildings.* Spon Press, New York.

The entropy law and the economic problem. In: *Energy and Economic Myths: institutional and analytical economic essays* (1976) Georgescu-Roegen, N. Pergamon Press, New York.

Change

The Fifth Discipline: the art and practice of the learning organization (1992) Senge, P.M. Random House, Sydney, Australia.

Mind and Nature: a necessary unity (1979) Bateson, G. E.P. Dutton, New York.

How Buildings Learn: what happens after they're built (1994) Brand, S. Penguin Books, New York.

Sustainability

World Commission on Environment and Development (WCED) (1987) *Our Common Future.* Oxford University Press, Oxford, UK.

Agenda 21: The Earth Summit Strategy to Save Our Planet (1992) Sitarz, D (ed.).

Cannibals With Forks: the triple bottom line of 21st century business (1998) Elkington, J. New Society Publishers, Gabriola Island, Canada.

Agenda 21 on Sustainable Construction (1999) International Council for Research and Innovation in Building and Construction (CIB). CIB Report Publication 237, July. Rotterdam, Netherlands.

Beyond Growth (1996) Daly, H. Beacon Press, Boston.

Small is Beautiful: a study of economics as if people mattered (1973) Schumacher, E.F. Vintage Books, London (reprint 1993).

Natural Capitalism: creating the next industrial revolution (1999) Hawken, P., Lovins, A., Lovins, L.H. Little, Brown & Co. New York.

Reflection time

Activity 1: Non-linear thinking

What is the link between wild salmon populations in local rivers and construction in Seattle, Washington, USA?

An independent organisation called Sustainable Seattle formed in 1993 to develop a set of indicators they could use to monitor the effects of urban development on the ecosystems within which Seattle is situated. One of the indicators they chose to adopt was the number of wild salmon present in local rivers.

Wild salmon have long been an important source of food and are of cultural significance to local Indian tribes. They have in the past also been an important commercial species. They require clean water and unobstructed runs in creeks and rivers as they return from the sea to reproduce. They have evolved to meet specific characteristics of their home stream water chemistry, including adapting their eggs to suit the predominant gravel size found in their home streams.

The salmon are therefore very vulnerable to the kinds of changes in water quality and landscape accompanying urban development. Many salmon runs have now been polluted by urban run-off or diverted by construction works.

Wild salmon populations in the north Puget Sound region have declined dramatically in the past 25 years. Sustainable Seattle's 1998 report states that salmon populations are dangerously low therefore marking Seattle's trend away from ecological sustainability. Chinook salmon is now a candidate for being listed as an endangered species.[1]

Discuss: How would a BEE change the way a building or housing subdivision in Seattle was designed and constructed to help restore salmon populations?

What is the link between mudslides, floods and construction sites in Malaysia?

'**Errant developers caused flooding – *Straits Times*, Singapore, 11 January 2000**
'KUALA LUMPUR – The government said it would come down hard on errant developers and contractors whose activities were largely responsible for the recent floods across the country.

'Deputy Prime Minister Datuk Seri Abdullah Badawi said irresponsible developers who clog up drains with mud and soil were the cause of recent floods in the Klang Valley.

'Floods can cause anguish to the people…Datuk Abdullah added that if the property developers had no sense of responsibility the people were the ones to suffer. Every time a flood occurred, the government had to spend money to tackle the problem…

'Heavy rains fell in Kuala Lumpur and parts of Selangor last Wednesday and caused floods in many low-lying areas…Thousands of motorists were caught in massive traffic jams when the main roads leading to Kuala Lumpur and Kalang were cut off by flood waters.'[2]

Straits Times, Singapore

Discuss: How would a BEE change the way a building or housing subdivision was designed and constructed to avoid blocking drains?

Activity 2: Imagine the future

'A sustainable world can never come into being if it cannot be envisioned. The vision must be built up from the contributions of many people before it is complete and compelling.'[3]

We all have different ideas about the way that we would like the world to be and there is no time but the present for us to consider how we might achieve our dreams. As we have read, the debate about how to achieve sustainability is problematic because many people use the same term to describe different futures. Sustainable development sounds environmental, but it might be financial. It could be ecological or social. Some say it is a dynamic balance between humanity's social, economic and ecological needs. Learning about sustainable development requires that you know and understand the various debates about what it is and how it can be achieved. But more fundamental to your personal and professional development is the need for you to decide what it means to you.

(a) Imagine your life 20 years from now and describe, using bullet points, the following:
  • Yourself
  • Your home
  • Your town
  • Your community
  • Your lifestyle
  • The government
  • How you interact with people

- How people interact with you
- Your workplace – what it is, where it is
- How you like to commute
- The food you eat
- The energy you use
- The natural environment

(b) With a partner choose key words that describe the conditions of the world that you each have described. Now make a list that synthesises you and your partner's preferred future conditions.

(c) Present this list to the others then synthesise all of the lists – what future conditions does the group share? What conditions are unique?

(d) With the group, make a list of how life is like NOW – Do you need to change your present in order to help the achieve your desired future? Circle on your 'NOW' chart or highlight areas requiring reform. Discuss how these reforms might take place.

(e) Which goals of sustainable development and sustainable construction presented in this chapter correlate with yours? Do you agree with the kinds of changes in current practice discussed in this chapter?

References

1 Sustainable Seattle Indicators Reports (1995, 1998) http://www.rri.org. Accessed 30/02/02.

2 *Straits Times* (2000) 'Errant developers caused flooding'. *The Straits Times* 11 January 2000, page 28, Mediacorp Publishing, Singapore.

3 Meadows, D., Meadows, D. & Randers, J. (1992) *Beyond the Limits: confronting global collapse, envisioning a sustainable future.* Chelsea Green Publishing Co. White River Junction, Vermont, USA.

PART III
THE BEEHIVE REVEALED

'Lovely
Through the paper window hole –
The galaxy.'

Issa[1]

With a basic understanding of the way nature works and the ways building professionals can act to accord with natural systems, a series of principles for ecologically sustainable building can be distilled. Preceding chapters have explained the scientific basis of these principles, so now they can be gathered together to provide a holistic picture of how ecologically sustainable building is created. The next sections reorganise the principles derived in the previous chapters into a series of nested project goals that apply to each phase in the building development life cycle. A case study is then presented to provide an example of how principles are holistically applied in the practice of creating ecologically sustainable buildings.

9 NATURAL LAWS AND PRINCIPLES OF ECOLOGICAL SUSTAINABILITY

Introduction

In this chapter we describe the difference between laws, principles and strategies for ecologically sustainable building. It is important to make a distinction between laws, principles and strategies because they represent different scales of influence over the ecological sustainability of a building project. We begin by discussing the difference between laws and principles. Then we gather the laws of ecological sustainability distilled from our investigation of thermodynamics and change and describe how the various principles help a building project adhere to these laws. Strategies describe actual techniques for enacting principles. Factors affecting the choice of strategy for a building project are then discussed. This leads us into a case study so we can see how all of this comes together.

Principles of action in building ecology either describe *laws* that set limits to what is possible, *principles* that describe the way things should be done, or *strategies* for taking action to heal destructive relationships with nature and to create healthy relationships. Principles and strategies only lead to ecology-sustaining outcomes when they operate within the limits set by laws. Principles* and strategies correspond to design while laws correspond to building process.[2] In nature all physical structures, from the tiniest cells to trees and mammals, are the result of genetic 'design', the process of creation, and the environmental factors that are acting on the process. These distinctions are important for understanding functions of design and process in nature and in building and therefore how and when to apply ecology-sustaining principles.

*Allen (2002) talks about 'Rules' rather than principles. Here the word 'principles' is used in place of 'rules' when referring to styles or intentions of action leading to ecology sustaining outcomes. This is because it is a more widely used and understood term.

Laws of ecologically sustainable building

Laws of nature provide constraints on what is possible in the construction and function of buildings. These constraints come into play when the building project takes the leap from paper design to physical structure.

The building process generally works at the limits of the constraints imposed on it by natural laws. For example, there is a maximum speed at which bricks can be laid that is a function of the weight of the brick, the height and complexity of a wall and the skill of the bricklayer. The geological constraints of a particular site impose a maximum bearing pressure that determines the size and configuration of footings. The difference between daytime and night-time temperatures in a location determines the effectiveness of *thermal mass* as a means of maintaining comfortable internal temperatures. The availability of raw materials determines the types of materials that the building can be constructed from.

Due to the organisation of finance, time and labour, building construction nearly always operates at the limits of what is possible under the laws that govern it. Unlike design, the building construction process causes a physical alteration in the environment that is impossible to change without creating further impacts. Constraints such as these should inform the building design process.

While the limits of construction possibilities imposed by nature cannot be overcome by design, they are not constraints on creativity. This is because the full range of construction possibilities given the constraints of natural laws is unknown. Architects might take comfort in visiting a rainforest. In one square kilometre there may be hundreds of different species of tree, there may even be hundreds of the same species tree, each with its own aesthetic, functional and structural quality. The design of trees in this small area of rainforest is constrained by the same natural laws, yet the variety of forms is staggering, as is the ecological integrity of the emerging system. While the structure and function of a building is constrained by natural laws, creativity in response is limitless.

The laws of thermodynamics are concerned with the conservation of material (matter), and the flow of energy. Natural systems have evolved to survive and prosper under the conditions that the laws of thermodynamics describe. The simple pattern of organisation that supports sustainable transfer of matter and energy, the hallmark of 'surviving designs', is the autocatalytic feedback loop.[3]

Ecologically sustainable building therefore must be considerate of the interactions between materials and energy and be organised for recycling. Based on these observations it is possible to describe a set of laws that govern ecologically sustainable building. In order to sustain ecosystems and accord with laws of thermodynamics, building must:

- consume resources no faster than the rate at which nature can replenish them (first law);
- create systems that consume maximum energy-quality (second law); and
- create and use byproducts that are nutrients or raw materials for resource production (fourth law).

It is important to emphasise that BEEs need to consider the effects of our decisions in relation to these laws over the entire building life cycle.

Principles for ecologically sustainable building

Principles for ecologically sustainable building provide the directions for our journey towards sustainability. The derivation of these principles is based on establishing ways building should be conducted so that laws of ecological sustainability governing nature's design process are adhered to.

Determining the way buildings should be built is a function of design. According to Pattee[4] and Allen,[5] rules about the way things should be done are derived from the observer's decisions about how to observe and what is significant. 'Rules [principles] are linguistic, local, arbitrary, and are rate-independent'.[6] While building design establishes rules for how a building will be built, design itself is not constrained by anything in nature. Designs can be changed almost at will and with minimal environmental impact in order to offer emergent possibilities for meeting a brief. How closely a completed building resembles, and operates in accordance with, its design is however still a function of how well a designer understands the constraints of the building process, the building environment, its occupants and its life cycle.

To use an example from nature, while a genetic code or 'design' describes what should be created, what is actually created is a variant. This is because the design is not strongly constrained by environment, while the development process is. For example, the structures of two genetically identical trees will differ according to the amount of light and nutrients, depth of soil and presence of competing organisms. The process of their growth is constrained by the limits imposed by their environment. In building, the design follows principles that describe the intended outcome; while the actual outcome is determined by the laws that govern the process of construction and the way people use the finished product.

Principles for ecologically sustainable building are of two types. Either they are principles for creating ecologically sustainable conditions, or they are principles for responding to existing ecologically unsustainable conditions. Ecology-'sustaining' principles will always be relevant because they respond to biophysical limits and ecosystems function – the constraints on process imposed by natural laws.

Ecologically sustainable building is a desired outcome of building design. We can therefore distil from the discussion in preceding chapters the principles for building that will lead us in that direction. Table 9.1 presents these principles 'nested' with the law that they assist in adhering to.

Table 9.1   Nested principles of ecologically sustainable building.

Law: Consume resources no faster than the rate at which nature can replenish them

Principles
Minimise resource consumption
Maximise use of renewable and used resources
Do more with less – resource efficiency

Law: Create systems that consume maximum energy-quality

Principles
Use solar income
Use energy in a large number of small steps, not in a small number of large steps
Minimise waste

Law: Create only byproducts that are nutrients or raw materials for resource production

Principles
Eliminating pollution
Use biodegradable materials before bio-accumulating materials
Reuse, then refurbishing and recycling of materials, components and buildings

Law: Enhance biological and functional adaptability and diversity

Principles
Apply life-cycle awareness and the precautionary principle
Provide access to fast-cycling materials without destroying slow-cycling materials
Protect and enhance biodiversity

Consume resources no faster than they can be replenished

This condition requires:

- minimising resource consumption;
- maximising use of renewable and used resources;
- do more with less – resource efficiency.

Minimising resource consumption

Kibert[7] describes the principle of minimising resource consumption as the 'conservation principle'. The intent is to encourage the conservation of resources in order to ensure that future generations have a stock of available assets from which to draw in order to fulfil needs. The use of 'conservation' widens the scope of this principle to include the concept of preservation, which engenders acts of protection and maintenance. Minimising resource use can, therefore, be more clearly seen as an instrument of resource conservation that is more applicable to finite resources required for physical production, including energy and land.

Ecologically sustainable buildings and building materials must require minimal energy for their manufacture, operation, maintenance and reuse. Ecologically sustainable building must be low in embodied energy and have low operational energy requirements. A building that is low in embodied and operational energy is said to have minimised life-cycle energy consumption.

Maximising use of renewable and used resources

In order to comply with first law conditions for ecological sustainability, building needs to shift from near total reliance on finite resources, to renewable resources. Renewable resources are solar energy and those terrestrial resources that are replenished at rates that allow for ecological well-being while satisfying human needs. Availability of renewable resources should not be thought of as a bottomless well. Renewables like timber, water and land need to be used in a sustainable manner. This requires that consumption of renewable resources only proceeds at rates that do not exceed an ecosystem's carrying capacity. Damage to these resources from pollution, over-consumption and mismanagement must be avoided so that nature can replenish them.

Reusing resources that are already in the system obviously reduces the consumption of raw materials from nature. Given that the building industry relies on vast quantities of finite raw materials, increasing the substitution of natural resources by mining the built environment for useful materials helps reduce environmental impacts and reduces loading on ecosystems as they work to replenish resource stocks.

Do more with less – resource efficiency

Resource efficiency means achieving more using less:

> 'Resource efficiency is the process of doing more with less – using fewer resources (or less scarce resources) to accomplish the same goals.'[8]

The construction industry is a major consumer of natural resources and therefore many of the initiatives pursued in order to create ecologically sustainable buildings are focusing on increasing the efficiency of resource use. The ways in which these efficiencies are sought are varied. Examples range from the principles of passive solar design, which aim to reduce the consumption of non-renewable resources for energy production, to life-cycle design and design for deconstruction. Methods for minimising material wastage during the construction process, and providing opportunities for recycling and reuse of building material and components, also contribute to improving resource efficiency.

Calls to be resource-efficient have been born from concern for the increasing depletion of non-renewable natural resources such as carbon-based fuels, minerals and forests. Minimising the waste of resources is also central to the concept. However, resource depletion can only be slowed and never eliminated by resource efficiency. This has led to some theorists and researchers[9,10] considering resource efficiency a conservative response to global environmental problems. How, for example, can any agency feasibly minimise the waste of an old-growth forest? It is impossible. In the case of forests, and any resource that it is important to 'conserve', being resource efficient without also using ecologically sustainable materials and designs will only serve to slow rates of depletion of the resource. This is of course a necessary step, but resource efficiency alone cannot lead to ecologically sustainable outcomes.

> 'Eco-efficiency – doing more with less – is an outwardly admirable concept. But it works within the industrial system that originally caused the problem. It presents little more than an illusion of change.'[11]

A simple technique for thinking about resource efficiency is to set ourselves the task of designing systems or choosing technologies that solve as many problems or perform as many functions as possible with the application of simple technology.

Create systems that consume maximum energy-quality

This condition requires:

- use of solar income;
- use of energy in a large number of small steps, not in a small number of large steps;
- minimising waste.

Use solar income

Like ecosystems, our modern economy and therefore our building industries are complex, far from equilibrium, systems that require the constant input of energy to maintain organisation. Unlike ecosystems, which are designed to turn sources of low-quality energy like sunlight into sources of high-quality energy like food and fuel, our economy and building industry turn sources of high-quality energy like coal, oil and materials into sources of low-quality energy like pollution and waste.

Nature's systems predominantly use continuously renewed solar income and thus maintain a high level of stability away from thermodynamic equilibrium. Ecologically sustainable building therefore must primarily use renewable energy derived from solar income rather than fossil fuels.

According to Odum,[12] consuming high-quality energy contributes to sustainability only if it 'interacts with and amplifies a lower quality, higher quantity energy source'. For the built environment this means using our current consumption of fossil fuels to drive the development and implementation of wide-scale use of renewable energy.

The advent of grid-connected photovoltaic systems in some countries is an example of our ability to make built environments that derive energy from solar income and, like ecosystems, feed it into the network as higher quality energy. Photovoltaic technology is not the only way to use solar income to create high-quality energy. Revegetation of urban areas and urban food production also make very effective use of solar energy, provide high-quality services to the community and help reduce fossil fuel consumption.

Use energy in a large number of small steps, not in a small number of large steps

Ecosystems evolve trophic levels representing the types of species and related food webs that exist to utilise energy of different qualities. These trophic levels are arranged in a hierarchy, from many species utilising low-quality highly abundant energy, up to only a small number of species that rely on energy of high quality and low abundance. The more mature an ecosystem is, the greater the number of steps there are in the trophic food chain.[13]

Better use of energy-quality is a key issue for BEEs. When they make decisions they always consider how completely energy-quality will be utilised as a result. Commonly, high-quality electrical energy is used only once to provide a simple service like incandescent lighting, yet much of the energy-quality left over after transformation of the energy into light escapes as diffuse heat and is not used – a waste of energy-quality.

While there has been a lot of improvement in energy efficiency, effective use of energy-quality has been neglected. However, there are good examples of strategies that improve the effectiveness of energy use. The use of heat exchangers to preheat or cool incoming air in *HVAC* systems, using thermal mass to store and radiate heat in a building, co-generation plants providing electricity and hot water and district heating and cooling systems that supply hot or cold water to a number of buildings in an area are all strategies for capturing and using energy of different qualities. Similarly, car-pooling makes better use of the energy-quality in our petrol, and reusing materials makes better use of their embodied energy qualities.

Minimising waste

BEEs avoid building construction and demolition waste because they understand that waste represents unused energy-quality and environmental impacts incurred for no benefit. Minimising waste can also significantly reduce disposal costs and therefore help make construction processes more profitable. In order for waste minimisation, or any environmental management programme, to be successful, it must involve and request commitment from all facets of the project team.[14,15] Some strategies for minimising construction and demolition waste are listed in Tables 9.2, 9.3 and 9.4.[16]

Table 9.2   Management responses for waste minimisation.[16]

Analyse project waste profile	Provide appropriate site transportation	Place onus for waste minimisation on subcontractors
Implement plan and document	Communicate and co-ordinate with clients and designers	Co-ordinate supervision of and communication with subcontractors
Cost control	Evaluate performance	Job-site separation of waste for recycling
Purchase material with minimal packaging, and which will not damage goods during delivery	Control purchasing to limit over-ordering	

Compiled from Graham, P. & Smithers, G. (1996) Construction waste minimisation for Australian residential development. *Asia Pacific Journal of Building & Construction Management* 2 (1) 15–16.

Table 9.3   Design responses for waste minimisation.[16]

Life-cycle flexibility	Retrofit and refurbish
Modular/prefabricated systems	Design for deconstruction
Specify recycled products	Detail construction documentation
Dimensions to suit standard material sizes	

Compiled from Graham, P. & Smithers, G. (1996) Construction waste minimisation for Australian residential development. *Asia Pacific Journal of Building & Construction Management* 2 (1) 15–16.

Table 9.4   Workers' responses for waste minimisation.[16]

Understand the consequences of waste
Build carefully to avoid waste
Look after materials from delivery to fixing
Store, stack and handle properly
Take responsibility for your own actions and waste
Participate in source separation activity

Compiled from Graham, P. & Smithers, G. (1996) Construction waste minimisation for Australian residential development. *Asia Pacific Journal of Building & Construction Management* 2 (1) 15–16.

Create and use byproducts that are nutrients or raw materials for resource production

This condition requires:

- eliminating pollution;
- use of biodegradable materials before bio-accumulating materials;
- reuse, then refurbishing and recycling of materials, components and buildings.

Eliminating pollution

The more complex a product is, the harder it is to recycle. More energy is required which means more polluting emissions to the environment. Materials containing toxic substances generally require the use and emission of toxic substances during their manufacture. The toxicity of the material and/or its production process can make recycling dangerous. Asbestos, for example, has proven to be carcinogenic during extraction, processing and incorporation into a building, and then poses no risk unless disturbed, normally during demolition.

Reducing the complexity and toxicity of building and building material design increases the ease of reuse, refurbishment and recycling and can decrease the energy required for reprocessing. This can decrease the cost of recycling and of used resources and, depending on the market for new materials, increase demand. Increasing demand for used materials provides an incentive for developing new infrastructure for recycling.

Use of biodegradable materials before bio-accumulating materials

Many building materials in common use are synthetic materials that nature cannot readily break down. Synthetic materials therefore become 'molecular garbage' rather than nutrients in a positive feedback cycle, building up in ecosystems, food chains and people.[17] In order to eliminate pollution a development must not only eliminate the release of toxic substances but also aid in the reduction of existing levels of contamination.

Biodegradable materials are any materials that can return to nature in a form that is at least benign and preferably a nutrient. Examples include earth and stone, timber and other untreated wood products, cellulose and natural fibres. If these materials cannot be reused within the built environment, then they have the potential to be useful to natural environments. If the byproducts of industry released into natural environments support ecological integrity, then ecosystems remain able to function and support human prosperity. If the outputs to nature of human activities in pursuit of prosperity support ecological prosperity, human well-being is also supported by the positive feedback loop created.

Reuse, then refurbishing and recycling of materials, components and buildings

This condition requires:

- reusing buildings and materials where possible;
- repairing and refurbishing materials rather than discarding them;
- recycling materials and using materials with recycled content.

According to the second law of thermodynamics energy is lost every time work is done. Therefore, the best use of energy-quality is achieved by creating feedback loops that allow for components from one building or construction process to be directly reused on either the same project or on another. The best feedback loops therefore are reuse loops. Direct reuse might apply to using existing buildings instead of building new, and reusing building components like windows or doors. Most building components and materials have to undergo reprocessing before they can be used again.

If the level of reprocessing does not change the nature or function of the original component then it can be regarded as refurbished. Feedback loops involving refurbishment of materials require more energy than reuse loops. A door that requires trimming, sanding and repainting has been refurbished prior to reuse, while the brass door furniture may be able to be taken from one door and reused on the refurbished door directly.

The lowest form of material feedback in this energy hierarchy is recycling. When a material is recycled it is broken down in some form of industrial process and becomes either raw material in the manufacture of the same recycled product, or a raw material in the manufacture of a different recycled content product. Different materials also require different amounts of energy to be recycled.

Ecologically sustainable building therefore creates and relies on post-consumer 'products' such as secondhand, recycled and recycled-content materials. In addition, sustainable buildings must only produce solid, liquid and gaseous emissions that can be used as 'food' for the production of new life-supporting goods and services. Creating demand for post-consumer products provides incentive for the establishment of more extensive and efficient recycling infrastructure. The better the recycling infrastructure, the easier it is to find and incorporate used materials, and the easier it is to find non-landfill destinations for leftover materials. A positive feedback cycle is therefore created.

Enhance biological and functional adaptability and diversity

This condition requires:

- apply life-cycle awareness and the precautionary principle;
- provide access to fast cycling materials without destroying slow cycling materials;
- protect and enhance biodiversity.

Apply life-cycle awareness and the precautionary principle

An ecologically sustainable approach requires making decisions that consider long-term as well as short-term issues, and which avoid environmental risks. These fundamental approaches require a life-cycle perspective and a precautionary approach to decision-making.

Protect and enhance biodiversity

Ecosystems are dynamic. The idea of sustaining an ecosystem therefore begs the question: in what state would it be sustained? Ecosystems vary in their complexity (number of interdependent parts or pathways) and their diversity (number of different kinds of interdependent parts or pathways). Assuming they do not suffer a major disturbance, ecosystems are thought to progress toward more complexity and diversity as they evolve new ways to degrade incoming solar energy. Sustaining ecosystems in a particular state is therefore not possible.

What is important is sustaining an ecosystem's functional stability, that is, its ability to develop complexity and diversity. It has been shown that, as ecosystems become more diverse, they develop resilience to change. They can therefore absorb greater magnitudes of change before losing their ability to function. The more resilient an ecosystem is the more stable and predictable is its behaviour. Because we depend on ecosystems to provide the goods and services we need to survive, it is important to sustain resilient ecosystems. The key to this is protecting their diversity and complexity. Damage has already been done to ecosystems due to rapid loss of biodiversity and complexity due to such human activities as logging and urban development. The imperative for ecologically sustainable buildings is therefore to protect and enhance biodiversity.

Provide access to fast cycling materials without destroying slow-cycling materials

The future is uncertain. Social and economic environments will change and in turn create demand for changing patterns of urban development and building requirements. Because buildings are predominantly constructed from finite materials, buildings constructed now must be designed to be adaptable in their structure and diverse in the range of potential use. In other words they must be able to be easily modified or deconstructed in order to provide resources for new building in the future. Providing the possibility of mining built environments rather than natural environments for building materials also reduces ecological impacts and impacts associated with the embodied energy of materials.

Strategies for ecologically sustainable building

Strategies for ecology-sustaining building assist building professionals to make decisions, and help buildings operate in ways that create ecologically sustainable outcomes. They describe specific building approaches that adhere to ecology-sustaining laws and principles. Strategies are constantly evolving in response to demands for information, accuracy and problem-solving by building professionals working on ecology-sustaining buildings. Their applicability or even success in leading to an ecologically sustainable outcome is contingent on many social, and economic factors as well as on the competency of the professionals involved in a project. Examples of common strategies are listed in this section. For explanation of the strategies refer to Chapter 11 (the case study or the 'more information' section for further reading) or refer to the glossary of strategies at the end of the book. To be successful, strategies must be appropriate to:

• social, cultural and economic contexts;
• the scale of the project and/or environmental problem or constraint;

215

- the development life-cycle phase to which they are applied.

Social and cultural contexts

Some strategies provide ways of adhering to natural laws while others provide ways of responding to ecologically damaging current practice in order to drive change toward sustainability. Because environmental conditions are never static, it stands to reason that strategies adopted by ecologically sustainable building projects need to change over time as conditions on the planet change and new priorities for survival emerge. Strategies must be appropriate to social, cultural and economic conditions. For example, ecologically sustainable strategies appropriate in a highly resource-consuming and waste-producing wealthy industrialised country will differ from appropriate responses required in poor countries where lack of infrastructure and poor access to resources in densely populated areas pose major health and ecological threats. These issues relate to the concept of appropriate technology.

Appropriate technologies are defined as 'the application of technological styles…that respect the natural ecosystems and local social and cultural patterns'.[18] Appropriate technology, therefore, is that which responds to the needs and capabilities of the community using it.[19] The concept of appropriate technology can be applied to strategies for ecology-sustaining building and as definition of an appropriate building project. In this case appropriate building projects are 'developments which are compatible with local human systems and technology'.[20]

For ecologically sustainable building, it is accepted that a technology is appropriate when it does not cause effects that undermine the ecological health of a particular region. An appropriate technology must also 'best fit' social and economic environments. The economic environment affecting a development proposal can be determined by drawing upon indicators such as local income, employment, population size, or the cost of finance. These indicators then become the basis for decisions about what is the best investment for a given situation.[21] These ideas can be distilled into a set of criteria that can be used to help decide on which approach to ecologically sustainable building is appropriate in a particular context. These are shown in Table 9.5.[18,22,23]

Table 9.5   Criteria for evaluating the appropriateness of technology.[18,22,23]

Systems autonomy

Can the technology be made, operated and maintained by the community rather than outside organisations or experts?

Harmfulness

The technology should do no long- or short-term harm to social or environmental conditions

Cost of technology

Both capital and on-going costs including life-cycle costs and environmental costs must be affordable to the community.

Wealth creation

Will the technology lead to the sustainable creation of wealth within the community and/or contribute to the community's immediate and long-term development needs?

Multiple benefits – simple solutions

Can the technology provide more than one benefit by solving many problems?

Adaptability

Can the technology be easily updated and modified as needs change or technology improves?

Collated from: Harding, R. (ed.) (1998) *Environmental Decision-making: the roles of scientists, engineers and the public*, Federation Press, Leichhardt, Australia, p.350; Carley, M. & Christie, I. (1993) *Managing Sustainable Development*, University of Minnesota Press, Minneapolis, MN, p.122; Bartelmus, P. (1986) *Environment and Development*, Allen and Unwin, Boston, MA.

Issues of scale

Strategies must also be appropriate to the scale of the problem or environment being considered. Some strategies may deal with global issues like climate change or ozone depletion; others may be specific to city, a neighbourhood, a site or a building. Some strategies may apply to processes for procuring buildings, some to innovation in the entire building industry. It is important to identify the scale of issues to which a building must be designed to respond. Ecologically sustainable building projects should stipulate design requirements for issues of different scale in the project brief. Strategies should be selected in order to address issues across scales where possible.

Making these decisions requires knowledge of the critical environmental issues arising at a particular place and time. While some critical issues are location-specific, others are regional or global issues

Table 9.6 An example of possible environmental issues at different scales for an Australian building project.[24]

Building	Local	Regional	Continental	Global
Energy use	Water scarcity	Logging of native forests	Greenhouse gas reduction	Global warming
Water use	Active community	Coastal water pollution	Loss of biodiversity	Ozone depletion
Wastes	Aboriginal heritage	Indigenous revegetation	Land-clearing	Ecosystem decline
Emissions	Health of a stream Car-based transportation	Cultural reconciliation	Resource consumption	

Source: Graham, P. (2000) *Bellbrae Primary School Redevelopment – Ecologically Sustainable Design Program*, October, Kensington Australia.

that always need to be addressed. For example, during the project master planning phase of a recent school redevelopment in Australia, the issues for different scales shown in Table 9.6[24] were identified in consultation with the client and incorporated into the ESD design brief for the project.

Development life-cycle phases – who does what?

Different strategies are required at different life-cycle phases of development. This is because decisions made at each phase of development have the ability to contribute different qualities of environmental performance to a building. Many potential impacts can be avoided if BEEs are aware of the best project stage to mitigate them. Experience on many building projects has shown that the earlier in a building process ecologically sustainable building issues are tackled, the better the results both for the environment and for the project budget.[25] BEEs especially know the importance of integrating ecological sustainability in planning and design phase because once the building begins to be constructed, it begins to draw on environmental resources and the environmental effects begin to accumulate. Once the building is operating, other effects associated with consumption of energy and materials, waste streams and occupant behaviour occur. Figure 9.1 represents the increased ability to affect a project's ecological sustainability if sustainability issues are considered early in the project life cycle.[26]

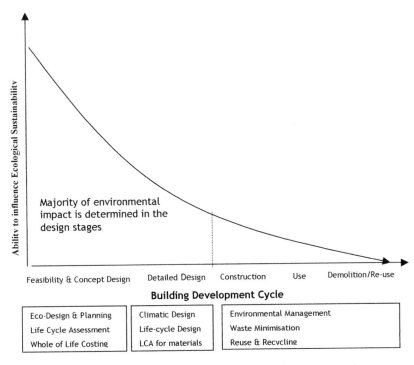

Examples of strategies available at each stage (there are many)

Fig. 9.1 The ability to create ecologically sustainable building is affected by when in the building process sustainability issues are considered. Reference: Lewis, H., Gertsakis, J., *et al.* (2001). *Design+Environment: a global guide to designing greener goods.* Greenleaf, London.

Feasibility phase

Traditionally the feasibility phase of a building project is concerned with determining financial costs and benefits and then creating financial relationships in order to secure the equity to develop. In addition to these financial issues, the feasibility phase of an ecologically sustainable building project requires determining the ecological costs and benefits of a project and understanding the relationships a development proposal will have with natural environments. In a financial context, the amount of equity available for a project provides constraints on development possibilities. In an ecological context, it is the carrying capacity of affected ecosystems that presents constraints.

During the feasibility phase of an ecologically sustainable building process, financial and environmental risks are assessed, and development decisions made based upon which deliver the best financial and ecological outcomes. To do this requires decision-making strategies that provide ways of determining both traditional financial and environmental relationships, and ways of integrating the financial and ecological effects of development proposals over their life cycle. Strategies for the feasibility phase of ecologically sustainable building include:

- whole of life costing;
- environmental impact assessment;
- environmental cost–benefit analysis.[27]

Development planning and design

Development plans and building designs can be blueprints for ecological disaster or sustainability depending on the effort taken during development planning and design to adhere to laws and rules of ecological sustainability. The role of development planning and design professionals is to avoid and minimise environmental risks posed by the development process and to specify ecologically sustainable building performance requirements. The intention to follow an ecologically sustainable development process must be described by project goals and incorporated into the design brief. The earlier ecologically sustainable goals and strategies are incorporated in the development process, the more successful they are.

Incorporating strategies in the development process

Development planning

- Bioregional planning
- Eco-city planning
- Habitat and species surveys
- Refurbish rather than build new

Building design

- Bio-climatic design
- Eco-design
- Long-life loose fit design
- Design for deconstruction

Construction, refurbishment and demolition

The ecological implications of a building begin to be felt once tenders are approved and construction begins. Construction, refurbishment and demolition occur throughout a building's life cycle. Environmental impacts are associated with the materials used, the amount of waste produced, the method of disposal, and the provisions in contracts for material salvage operations. The other major impact of construction is disruption to the site and surrounding areas during construction. On ecologically sustainable projects construction and demolition contractors adopt strategies that minimise the impact of their work processes and that ensure the integrity of ecologically sustainable aspects of building design. Appropriate strategies for the construction, refurbishment and demolition phases include:

- minimise material transportation to site and on-site;
- purchasing timber from certified sustainable sources;
- implementing environmental management systems on site;
- implement construction waste minimisation systems.

Operation phase

This is the longest phase in most building life cycles. During this phase the ecologically sustainable features of a design should begin to have a positive effect on environments over time. The most important contributors to the environmental effect of a building are the people who occupy it. Unless a building's ecologically sustainable features are used properly, many of its environmental performance attributes will remain unrealised. The occupants of buildings must therefore know how to optimise the environmental performance of their building. They also need management processes to maintain adherence to ecologically sustainable rules, and feedback about how the building is actually operating. Ecologically sustainable strategies for this phase therefore fall into the categories of instructions, management, and feedback.

Instructions

Cars come with owner manuals, VCRs come with instructions, and so should ecologically sustainable buildings. Designers and builders of ecologically sustainable buildings provide data not only on how equipment works but how the building should be operated as a system in order to maximise environmental benefits. Instructions should cover the use of building systems like ventilation, cooling and heating systems, as well as how to use the structure of the building and the properties of the materials that it contains. Information about the structure of a building is especially important when it has been designed for deconstruction because contractors need to know how to remove building elements non-destructively.

Management

While designers and builders have a large influence over the physical structure and form of a building, they have no control over many of the material inputs throughout its life cycle. Consumables like light fittings, computer cables, air diffusers, cleaning products, office sup-plies, even carpets and partitioning, are added and removed from a building in the course of normal operation and periodic maintenance. In ecologically sustainable buildings the selection, installation and disposal of these materials must adhere to the same ecologically sustainable rules that govern building design and construction. The responsibility for this normally rests with building owners and man-agers.

Feedback

Feedback is an essential ingredient for change. As we discussed in Chapter 7, receiving accurate feedback about the effects of our actions is necessary for us to alter our behaviour to optimise living condi-tions. Building occupants need to be provided with ways of monitor-ing their building's environmental performance over time, and with simple approaches to making appropriate adjustments. Buildings can also provide feedback about impacts and issues associated with the construction of the building and the reasons why certain materials were chosen. By incorporating feedback mechanisms a building can not only help occupants make it work best for the environment, it can teach them about the environment, their relationship with it, and the impacts their decisions have on it.

Strategies must be selected to best address sustainability issues relevant to a particular project. One way to determine what strategy a building project needs to incorporate to contribute to ecological sustainability is to use the ecologically sustainable principles and laws listed above as a checklist. Ask, for example, will this project or decision eliminate pollution? If not, then a strategy needs to chosen that allows the project or decision to adhere to this principle. In this case appropriate strategies might be to provide on-site treatment and reuse of wastewater, provide convenient transport alternatives to cars, or to find alternatives to toxic materials.

Incorporating strategies in the operation phase

Instructions

- Operating and maintenance instructions for the building as a system
- Deconstruction instructions
- Information on material properties including maintenance, reuse and recycling

Management

- Building environmental management planning
- Ecologically sustainable material purchasing policies
- Waste minimisation planning

Feedback

- Make environmental performance indicators like energy and water consumption, material life cycle impact data, and waste reduction visible
- Set environmental targets, monitor performance and share results
- Use building environmental performance rating schemes to promote good practice
- Expose resource supply and disposal pathways

It is important to remember that principles for ecologically sustainable building flow from ecologically sustainable laws. They help identify problems with building design, construction or operation and describe goals for ecological sustainability that the project should achieve. Once principles for the project have been chosen it is time to take action by specifying or applying strategies. Table 9.7 is by no means a comprehensive list of strategies for ecologically sustainable building, however, it shows how laws, principles and strategies can be 'nested' in order to provide a framework for implementing an ecologically sustainable building process.

Table 9.7 Nested strategies for ecologically sustainable building.

Create systems that consume maximum energy quality

Principle no. 1	**Decision-making phase**
Use low-quality, highly abundant energy resources	
Strategies	
Use retail renewable energy	Feasibility
Generate electricity using solar or wind energy	Feasibility
Create community gardens for local food production	Development planning
Passive climatic design	Design
Insulate to capture latent heat for occupant comfort	Design
Use stable soil temperatures as a passive heat exchanger	Design

Principle no. 2
Use energy in a large number of small steps, not in a small
number of large steps

Strategies	
Use heat exchangers to pre-heat or pre-cool incoming air or water	Design
Use materials that can be easily repaired rather than disposable materials	Design
Use non-composite materials	Design
Long-life loose fit design	Design
Provide easy access for replacement/repair of building components that wear at different rates	Design

Principle no. 3
Minimise waste

Strategies	
Reuse grey and black water	Development planning
Design for deconstruction	Design
Prefabricated and modular design	Design
Use materials with small ecological rucksacks	Design
Construction waste minimisation as a condition of tender	Design
Job site waste separation for reuse and recycling	Construction
Refuse over-packaged materials	Construction

Table 9.7 *(Continued.)*

Consume resources no faster than the rate at which nature can replenish them

---

Principle no. 1	**Decision-making phase**
Maximise use of renewable and used resources	

Strategies	
Use treated grey and black water for non-potable purposes	Development planning
Passive climatic design	Design
Use solar hot water	Design
Use materials and products that are easily recyclable	Design
Use secondhand or recycled materials	Design/construction

Principle no. 2
Minimise use of new and non-renewable resources

Strategies	
Refurbish existing buildings before building new	Feasibility
Reduce car-based transportation	Development planning
Choose low embodied energy materials	Design
Use secondhand or recycled materials	Design/construction
Construction waste minimisation as a condition of tender	Design/construction
Design for life cycle	Design/construction

Principle no. 3
Do more with less

Strategies	
Use lightweight high-strength materials in structural design	Design
Use tension before compression in structural design	Design
Choose technology that serves more than one purpose	Design
Minimise mechanical heating, ventilation and cooling	Design
Use energy and water efficient appliances and fittings	Design
Insulate existing buildings to reduce energy consumption	Design

Table 9.7 *(Continued.)*

Create only byproducts that are nutrients or raw materials for resource production

Principle no. 1 Eliminate pollution	**Decision-making phase**
Strategies	
Capture, treat and reuse waste water	Development planning/design
Reduce car-based transport	Development planning/operation
Passive climatic design	Design
Use materials that require minimum transportation	Design/construction/operation
Implement environmental management procedures during construction	Construction

Principle no. 2 Use biodegradable materials before bio-accumulating materials	
Strategies	
Use life-cycle assessment to choose materials that:	Design
• don't contain, create, use or emit toxic substances over their life span	
• require minimal processing, transportation, and maintenance	
• can be readily reused and/or recycled	
• don't contain substances that are accumulating in the biosphere	
Avoid using synthetic and toxic materials	Design/construction
Reuse and recycle synthetic and toxic materials	Design/construction/operation

Principle no. 3 Reuse, then refurbish and recycle materials, components and buildings	
Strategies	
Long-life loose fit design	Design
Design for deconstruction	Design
Use secondhand or recycled materials	Design/construction /operation
Construction waste minimisation as a condition of tender	Design/construction
Create time provision in demolition contracts for material salvage	Construction
Job site separation of construction waste for recycling	Construction

Table 9.7 (*Continued.*)

Enhance biological and functional adaptability and diversity	
Principle no. 1 Apply life-cycle awareness and the precautionary principle	**Decision-making phase**
Strategies	
Set environmental performance targets and monitor actual vs. predicted performance	All
Use LCA tools to select environmentally friendly materials	Design
Use energy modelling tools to predict operational energy demands and environmental impacts	Design
Use environmental design tools and information to avoid making decisions with uncertain environmental outcomes	Design
Principle no. 2 Provide access to fast-cycling materials without destroying slow-cycling materials	
Strategies	
Long-life, loose fit design	Design
Design for deconstruction	Design
Avoid composite materials, especially for finishes	Design
Principle no. 3 Protect and enhance biodiversity	
Strategies	
Build on previously developed land	Feasibility/development planning
Conduct habitat and species surveys during development planning	Development planning
Bioregional planning	Development planning
Contain urban sprawl	Development planning
Provide areas for local organic food production	Development planning
Indigenous landscaping and revegetation	Development planning/design
Don't use rainforest or old-growth forest timber	Design/construction
Eliminate site run-off during construction	Construction

References

1 Eighteenth-century haiku poet. Source: Donegan, P. (1990) Haiku and the Ecotastrophe. In: *Dharma Gaia – A Harvest of Essays in Buddhism and Ecology.* Parallax Press, Berkeley, CA, p.201.

2 Allen, T. (2002) Applying the principles of ecological emergence to building design and construction. In: Kibert, J., Sendzimir, J. & Guy, B. (eds) *Construction Ecology: Nature as the basis for green buildings.* Spon Press, New York. pp.108–126.

3 Odum, H. (2002) Material circulation, energy hierarchy, and building construction. In: Kibert, J., Sendzimir, J. & Guy, B. (eds) *Construction Ecology: Nature as the basis for green buildings.* Spon Press, New York. pp.35–71.

4 Pattee, H. (1978) The complementarity principle in biological and social structures. *Journal of Social Biological Structures* **1** 191–200.

5 Allen, T. (2002) Applying the principles of ecological emergence to building design and construction. In: Kibert, J., Sendzimir, J. & Guy, B. (eds) *Construction Ecology: Nature as the basis for green buildings.* Spon Press, New York. pp.108–126.

6 Allen, T. (2002) Applying the principles of ecological emergence to building design and construction. In: Kibert, J., Sendzimir, J. & Guy, B. (eds) *Construction Ecology: Nature as the basis for green buildings.* Spon Press, New York. p.9.

7 Kibert, C.J. (1994) Establishing principles and a model for sustainable construction. *Sustainable Construction.* Proceedings of the First International Conference of TG16. 6–9 November, Tampa Florida, USA.

8 Wilson, A., Uncapher, J., McManigal, L., Lovins, L.H. & Hunter, L. (1998) *Green Development: integrating ecology and real estate.* John Wiley, New York. p.7.

9 McDonough, B. & Braungart, M. (1998) The Next Industrial Revolution. *The Atlantic Monthly,* October.

10 Hawkin, P. (1993) *The Ecology of Commerce.* Harper Collins, New York.

11 McDonough, B. & Braungart, M. (1998) The Next Industrial Revolution. *The Atlantic Monthly,* October. p.62.

12 Odum, H.T. (1996) *Environmental Accounting: EMERGY and environmental decision making.* John Wiley & Sons, New York. p.26.

13 Schneider, E. & Kay, J. (1994) Life as a manifestation of the second law of thermodynamics. *Mathematical Computer Modelling* **19** (6-8). Elsevier Science, Amsterdam, Netherlands.

14 Lingard, H., Smithers, G. & Graham, P. (2001) Improving solid waste reduction and recycling performance using goal setting and feedback. *Construction Management & Economics* **19** 809–817.

15 Griffith, A. (1994) *Environmental Management in Construction.* The Macmillan Press Ltd, London, UK.

16 Graham, P. & Smithers, G. (1996) Construction waste minimisation for Australian residential development. *Asia Pacific Journal of Building & Construction Management* **2** (1) 15–16.

17 Robert, K.H. (1993) Answering the King's challenge, news from Sweden's Natural Step. *Designing a sustainable future, IN CONTEXT, a quarterly of humane sustainable culture* **35** (Spring) http://www.context.org/ICLIB/IC35/Robert.htm. Accessed 24 August 2000.

18 Bartelmus, P. (1986) *Environment and Development.* Allen & Unwin, Boston, USA.

19 Milbrath, L.W. (1989) *Envisioning a Sustainable Society: learning our way out.* State University of New York Press.

20 Yap, N. (1989) *Sustainable Community Development: an introductory guide.* Ontario Environment Network, Canada. Cited in Hill, R.C. & Bowen, P.A. (1997) Sustainable construction: principles and a framework for attainment. *Construction Management and Economics* **15**, 223–239.

21 Harvey, J. (1981) *The Economics of Real Property.* Macmillan Publishers Ltd, Hampshire, UK.

22 Harding, R. (ed.) (1998) *Environmental Decision-making: the roles of scientists, engineers and the public.* Federation Press, Leichhardt, Australia, p.350.

23 Carley, M. & Christie, I. (1993) *Managing Sustainable Development.* University of Minnesota Press, Minneapolis, MN, p.122.

24 Graham, P. (2000) *Bellbrae Primary School Redevelopment: ecologically sustainable design program*, October. Kensington, Australia.

25 Wilson, A., Uncapher, J., McManigal, L., Lovins, L.H., Cureton, M. & Browning, W. (1998) *Green Development: integrating ecology and real estate.* John Wiley & Sons, New York.

26 Lewis, H., Gertsakis, J. *el al.* (2001). *Design+Environment: a global guide to designing greener goods.* Greenleaf, London, UK.

27 Langston, C. & Ding, G. (eds) (2001) *Sustainable Practices in the Built Environment.* Second edition. Butterworth Heinemann, Oxford, UK. pp.61–84.

10 DEVELOPING ECOLOGICAL SUSTAINABILITY IN BUILT ENVIRONMENTS

Ecological sustainability is an attribute of the whole system of which a building is a part. Because of its life-cycle metabolism a building cannot be ecologically sustainable by itself. As such the preceding conditions for ecologically sustainable building derived from laws of thermodynamics apply to any scale of the built environment. Because of this we need to know whether applications of these laws and principles to our building are affecting the kinds of changes required to develop ecological sustainability in our built environment.

Thermodynamics in general, and particularly the second law, has been used to understand how and why ecosystems develop the way they do. One property of ecosystems is that they develop in ways that make best use of available energy-quality[1-3] and in so doing begin to exhibit attributes like functional stability and resilience that we strive to attain in the development of 'sustainable' built systems. In this section we will use an ecological understanding of the effects of thermodynamics on ecosystems to derive some indicators for sustainability in our built environments.

How should we progress?

The laws of thermodynamics discussed in this chapter describe fundamental physical conditions that life on Earth has evolved to cope with. The conditions described by these laws have also shaped the non-living elements of our world, therefore creating the basic constraints within which we live our lives. Nature's response to these conditions has been to develop systems that do more with less, best use energy-quality, and reinforce the production of the energy they need by feeding back outputs of consumption into food for production. As Schneider and Kay[4] summarise, 'ecosystems develop in a way which systematically increases their ability to degrade [use] the incoming solar energy'. Ecologists have noticed that these qualities help ecosystems become more resilient over time by increasing the systems complexity and diversity, and in so doing, its maturity and ability to survive.[5]

By applying new understanding of the implications of thermodynamics in living systems, ecologists have been able to create a description of the properties of ecosystems that are developing toward maturity.[6] The properties of resilience, functional stability and survival in maturing ecosystems are certainly components of the concept of sustainability that we strive to engender in our built environments. The description of these properties can therefore be used as a model for the properties of built environments that are developing ecological sustainability.

Properties of ecosystem maturity and built-environment sustainability

If built environments are to work the same way as natural environments then they must exhibit similar characteristics to the ecosystems of which they are a part. Ecosystems are, however, dynamic systems and exhibit different rates of change and stability at different phases in their development. Ecosystems become less unpredictable and more stable as they develop maturity.[7] According to Schneider and Kay, ecosystems develop maturity via systematic increases in their ability to use incoming solar energy.

Mature ecosystems exhibit characteristics of functional stability, organisation, diversity, efficiency, and adaptability to change. Ecosystems attain maturity over time in a successional process of development. These qualities of mature ecosystems also describe our functional aspirations for ecologically sustainable development in built environments. The environmental effects of our consumption of fossil fuels, for example, certainly indicate a need to systematically increase our built environment's ability to use incoming solar energy. The process of developing functional sustainability in built environments is therefore analogous with the development of maturity in ecosystems.

A distinction needs to be made here between functional aspirations and economic, social and cultural aspirations. Thermodynamics explains the basic conditions within which our systems are created. Designing systems that accord with thermodynamic conditions creates the supporting framework, the skeleton if you like, upon which our social, cultural and economic aspirations of sustainability can grow.

231

If we don't have systems that function thermodynamically in ways that support the development of sustainability then any social, economic or cultural aspirations are hobbled. That said, it is important to realise that social, economic and cultural conditions create powerful feedback that profoundly affect the decisions we make when designing our systems. They are therefore integral to a holistic approach to developing sustainability. The following system conditions, based on the work of Schneider and Kay, describe the functional attributes of an ecosystem that is developing maturity. Extending the analogy between achieving maturity and developing ecological sustainability, we can use these attributes as indicators of developing sustainability in built environments. Therefore, the process of developing sustainability in built systems and maturity in ecosystems would result in systems with:

- more energy capture;
- more energy flow activity within the system;
- more cycling of energy and material;
- higher average trophic structure;
- higher respiration and transpiration;
- larger ecosystem biomass;
- more types of organisms (higher diversity).

More energy capture

Ecosystems develop structure and organisation in order to degrade energy; the more energy that flows into a system, the greater the potential for energy degradation. In the built environment, structure and organisation are made possible by inflow of capital. The more capital available for building, the more can be built. In short, we can't have a built environment that develops sustainability unless we first have enough capital to create a built environment.

Maturing ecosystems are thought to systematically increase their ability to capture energy. Ecosystems rely on solar energy and develop many ways of capturing energy that is flowing into their physical location. Built environments, on the other hand, at present rely on developing better ways of capturing and importing energy from remote locations. Doing this increases disorder in the natural systems that are providing the energy and increases the financial cost of extending infrastructure. The continuing degradation of ecosystems all over the planet is an obvious indicator of this increasing disorder.[8]

To exhibit similar qualities to a maturing ecosystem, built environments must rely less on energy imports and develop ways of capturing more of the energy flowing naturally into their physical location. Indicators of progress toward sustainability for a built environment could therefore include the level of renewable energy captured and used by a built environment as a percentage of imports of non-renewable energy used. Use of photovoltaic, wind and co-generation energy would be prevalent in a built environment that is developing sustainability. You might also expect to see more plants and animals, and local food production – anything that captures energy that is naturally available locally.

More energy flow activity within the system

The rationale is the same as the above condition – the more energy there is flowing, the more energy there is potentially available for use. The key distinction here that energy flow *within* the system is increasing. This condition relates to the requirement to effectively utilise energy of different qualities. If the amount of energy that a system receives is finite (which is the case in nature) then energy flow within the system needs to be enhanced to maintain it. For built environments this means that energy in all its forms must be used effectively and that it must be distributed throughout the system.

More cycling of energy and material

Number of cycles increases

The more pathways that exist for energy to be recycled within the system, the more completely incoming energy is used. For the built environment the implications are obvious. Rates of materials and energy reuse and recycling should be maximised. The variety of materials and energy that are reused and/or recycled should also be maximised. This in turn requires increasing the availability of recycling infrastructure.

The length of cycles increases

The more mature an ecosystem, the longer the cycle. This means that there are more nodes in the cycle. This might mean, for example, that all buildings are designed for deconstruction in order to first gain access to building components and systems for reuse. Later in time the component materials might be removed and reused individually. Once their utility as a material is diminished, they may become feedstock for the production of new materials either within the building industry or beyond it. In this way embodied energy-quality of materials is more completely utilised. Indicators might include the prevalence of deconstruction rather than demolition, and the number of buildings that are using passive climatic systems to provide indoor comfort.

From an operational energy perspective this condition would mean that energy entering a building was always put to its highest and best use first, before being captured and fully utilised again in different forms as it degrades. The imperative is to consume maximum energy-quality.

The amount of material flowing in cycles (as opposed to straight through) increases

This condition shows that less energy and materials are being lost from the system – the system becomes less 'leaky'. Under these conditions the amount of resources retained in useful service is maximised and waste is minimised. This condition could be indicated by decreasing rates of solid waste going to landfill or incinerators, and the volume of building materials reused and recycled.

Turnover time of cycles or cycling rate decreases

More nodes in the system mean that there are more places for material to be stored. Cycling of materials through the system is therefore slower. In mature ecosystems more material is stored in biomass than is in production. The implication for the built environment of this condition is that as it develops sustainability, more material is stored or cycling within the system than is imported to it. It therefore accumulates a built-environment version of biomass. This would be indicated by increased use of existing buildings and refurbishment, which implies buildings designed for greater functional adaptability, increased reuse of materials, increased durability of materials, and increased material salvage and resale.

Higher average trophic structure

Longer trophic food chains

In food chains energy is degraded at each step in the chain. The more species there are involved in a food chain, the longer it is and the more completely available energy can be used. This condition highlights the importance of increasing the diversity of producers and consumers. In a built environment developing sustainability we might expect to see an increased diversity in materials and building types as producers and consumers fully exploit niches in the market. In developed urban environments, we might see the increased use of low embodied energy materials and so-called 'alternative materials' like straw bales (see Fig. 10.1), rammed earth and mud brick and different styles of development including community title might gain wider acceptance. In developing communities we might see development of improved designs of indigenous building techniques. Of course this is conditional on economic and regulatory parameters being set to allow for this diversity to exist.

Species will occupy higher average trophic levels

Energy at higher trophic levels is of higher quality. Because ecosystems evolve to best use energy-quality, species in mature ecosystems have the capacity to consume higher quality renewable energy sources available within the system. More energy-quality is consumed and less wasted. Surface temperatures above mature ecosystems have been recorded as cooler than temperatures above less complex ecosystems and land areas, indicating that less usable energy escapes a mature system.[9]

In a built environment that is developing sustainability therefore there would be an increased use of high-quality renewable energy. More local food production, increased use and efficiency of solar electricity and hot water and more absorption of solar energy from plants, including on roof and wall areas, might indicate this. We might expect to see far more integration of habitat in urban areas and more built space being used to support green space. Indicators might include reduction in the heat-island effect due to decreased radiant heat from urban areas, increasing number of plants per square metre (both horizontal and vertical area), number of appropriately insulated buildings and the area of urban form used for gardens and food production.

Fig. 10.1 Increasing availability and use of alternative building materials is an indication of developing sustainability. This straw-bale cottage is one of four being constructed alongside concrete block apartments. The buildings are part of the 'Christies Walk' Eco-City development in inner-city Adelaide developed in association with the UN registered non-government organisation, Urban Ecology Australia.

Greater trophic efficiencies

More of the low-quality energy that an ecosystem receives is turned into higher-quality energy. In the built environment this is indicated by increased efficiency with which renewable energy is captured and converted into high-quality energy. Decreasing cost of renewable energy supply and the increasing efficiency of production might indicate this.

Higher respiration and transpiration

Respiration and transpiration are the processes by which oxygen is absorbed and carbon dioxide released through foliage. If we imagined a city's roofs as the canopy foliage of a forest, then we would expect roof space in a built environment developing sustainability to be increasingly becoming garden space (see Fig. 10.2). More integration of habitat and more trees in urban areas would increase rates of respiration and transpiration.

Respiration and transpiration can also be thought of as breathing in and breathing out. Buildings must also 'breathe'. Unfortunately, particularly in the commercial sector, buildings have been designed like iron lungs, requiring huge quantities of electrical energy to move air in and out of their sealed facades. Naturally ventilated buildings on the other hand are designed to breathe without the use of high-quality imported, non-renewable energy. Instead they are designed to integrate with local climatic conditions, better utilising the quality of locally available solar energy to maintain healthy environments. Most buildings in a built environment developing sustainability would therefore be designed for their local climate, infused with and taking advantage of locally available solar energy; breathing without a respirator.

Larger ecosystem biomass

Mature ecosystems have more biomass. Therefore in a built environment developing sustainability there would be more organic material, both alive and dead, in the system. Indicators might be increased habitat areas, and high (if not complete) rates of organic waste composting and recycling. There may be more variety of organic waste management and biological treatment including the use of biological systems to replace mechanical systems for waste treatment and disposal. We might also notice increasing areas of indigenous habitat, and increasing rates of composting of organic material.

More types of organisms (higher diversity)

From an ecological perspective increased species diversity provides the system with resilience and adaptability, increasing its resilience in the face of change.[10]

Fig. 10.2 A green idea – the hanging gardens of Melbourne? Increasing the density of urban vegetation increases evaporation and transpiration. This image is a conceptual vision of the central business district of Melbourne, Australia, having being retrofitted with plants using the Green Wall™ suspension system. The Green Wall system provides a lightweight structural steel framework onto which different species of plants are grown. The developers of the system hope to reduce heat load on building facades and increase inner-city air quality by adding plants to the sides of existing buildings. Image courtesy of Boughouse Pty Ltd.

In the built environment we may see increased species diversity through increasing the infusion of more mature ecosystems into urban environments. These might be evident by establishment of wildlife corridors, de-channelisation of creeks and rivers, and urban revegetation projects. The city then becomes a home for a greater variety of non-human life, and provides new pathways for utilisation of solar energy. Considering non-living components of the built environment, diversity applies to the utility of building structures, variety of building types, and transport options. In a built environment developing sustainability, citizens have more choice and mechanisms for involvement in shaping their environments to adapt to change. Increasing participation in and success of habitat regeneration programmes and increasing number and variety of species might indicate that higher diversity within urban areas was being achieved.

Beyond the urban boundaries, diversity would also be increasing in ecosystems that the built environment is interrelated with, as a result of investment in habitat restoration and protection, and reduced loading as a result of decreases in the exporting of pollution and reliance on ecosystems as waste sinks. The Sustainable Seattle program, an urban sustainability initiative in Washington State, USA, as we have discussed, counts the number of salmon in local rivers as an indicator of wider ecosystem health resulting from the city's progress.[11]

These indicators only deal with the physical environment of an urban area developing sustainability. There are also, as principles of sustainable development and construction describe, social, cultural and economic prerequisites for this style of change.

References

1 Schneider, E. & Kay, J. (1994) Life as a manifestation of the second law of thermodynamics. *Mathematical Computer Modelling* **19** (6-8). Elsevier Science, Amsterdam, Netherlands.

2 Kay, J. (2002) On complexity theory, exergy and industrial ecology. In: Kibert, J., Sendzimir, J. & Guy, B. (eds) *Construction Ecology: Nature as the basis for green buildings*. Spon Press, New York. pp.72–107.

3 Odum, H. (2002) Material circulation, energy hierarchy, and building construction. In: Kibert, J., Sendzimir, J. & Guy, B. (eds) *Construction Ecology: Nature as the basis for green buildings*. Spon Press, New York.

4 Schneider, E. & Kay, J. (1994) Life as a manifestation of the second law of thermodynamics. *Mathematical Computer Modelling* **19** (6–8). Elsevier Science, Amsterdam, Netherlands. p.38.

5 Odum, H.T. (1996) *Environmental accounting: EMERGY and environmental decision making*. John Wiley & Sons, New York.

6 Schneider, E. & Kay, J. (1994) Life as a manifestation of the second law of thermodynamics. *Mathematical Computer Modelling* **19** (6-8). Elsevier Science, Amsterdam, Netherlands.

7 Peterson, G., Allen, C. & Holling, C. (1998) Ecological Resilience, Biodiversity, and Scale. *Ecosystems* **1**, 6–18. New York, Inc. USA.

8 United Nations Development Program (UNDP), United Nations Environment Program (UNEP), World Bank & World Resources Institute (2000) *A Guide to World Resources 2000–2001: people and ecosystems: the fraying web of life.* World Resources Institute, Washington DC, USA.

9 Luvall, J. & Holbo, H. (1989) Modelling surface temperature distributions in forest landscapes. *Remote Sensing and Environment* **27** 11–24.

10 Peterson, G., Allen, C. & Holling, C. (1998) Ecological Resilience, Biodiversity, and Scale. *Ecosystems* **1**, 6–18. New York, Inc. USA.

11 Sustainable Seattle Indicators Reports (1995, 1998). http://www.rri.org. Accessed 30/02/02.

11 CASE STUDY

Introduction

The purpose of this chapter is to provide an example of how laws, rules and strategies of ecologically sustainable building have been applied in practice. The case study is the Ostratorn School in Lund, Sweden. This school was designed and built using 'eco-cycles', a method that considers the relationships between the building and the social, economic and natural environment at each of its life-cycle phases. The eco-cycles approach provides a holistic systems-based method for ensuring that the project adheres to laws and principles of ecologically sustainable building, and takes advantage of the effectiveness of natural systems.

Case study: Ostratornskolan, Lund, Sweden

'A school built on ecological ideas'

Project information

Name: Ostratorn Skolan
Location: Lund, Sweden
Brief: Upgrade and extend an existing school
Client: Lundafastigheter
Architect: White Arkitekter AB
Completed: October 1998

Background

The project at Ostratorn School was to refurbish and extend existing school buildings and construct two new classroom blocks. Construction was completed in 1998. The project is a great example of ecologically sustainable building. Its design, construction and operation recognise that nature works in cycles and not straight lines. The buildings work with biogeochemical cycles to avoid disrupting them, rely on solar income and passive energy flows more than fossil fuels, and make use of materials from built environments rather than from natural environments.

The Ostratorn School in Lund, Sweden, has been designed and constructed with natural cycles in mind. The design thinking that created this school considers the origin, use and destination of materials, water, energy and food. The buildings connect with the site, its climate and the local economy is considered through 'eco-cycles' which acknowledge the interdependencies of the school with its environment. Eco-cycles also provide a way of thinking that closes loops rather than leaves open pathways for waste, pollution and over-consumption of resources.

The concept of eco-cycles revolves around a consciousness of the cycling of resources, air, heat and water throughout buildings and their constituent life cycles. It recognises that an ecologically sustainable building must be not only efficient but also effective with the resources it uses.

Resources

Resources required for the school were categorised as food, materials, water and energy. An 'eco-cycle' for each resource was designed in order to understand how resources could be kept in service within the school and the local community.

Recognising that construction requires the use of many non-renewable and energy-intensive materials, the designers asked themselves: How can we use as few non-renewable resources as possible to build the school? They asked: Which materials and construction techniques can be recycled by nature or the community when the building is renovated or demolished?.[1]

Recognising that waterways are often overburdened with nitrates and phosphates from sewerage system discharge and urban run-off, the designers asked themselves: Would it not be better to use the nutrients in urine as fertiliser on fields, than to burden the sewer system?

They also realised that separating urine and faeces made it easier to recycle solid waste as fertiliser for crops that provide the school with food.

Recognising that all forms of large-scale power generation have environmental disadvantages, the designers required that as little electricity as possible should be used. They asked: How can the need for energy, primarily electricity for heating and operation, be kept as low as possible? How can we utilise renewable, local sources of energy? Mapping eco-cycles for each resource was the key to creating a project that answered these questions.

The food eco-cycle

The food eco-cycle (Fig. 11.1) explains how food and food-related waste is reused and recycled to continue to help feed the school and reduce reliance on outside sources. Food is delivered to the school and prepared as meals. Non-biodegradable materials used in cooking and

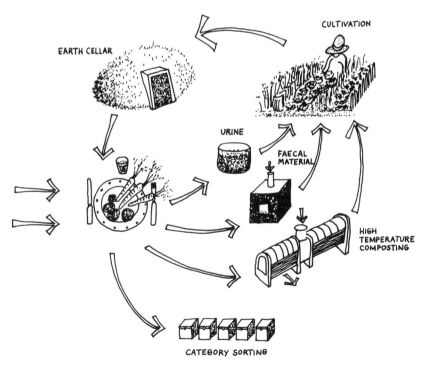

Fig. 11.1 Food eco-cycle. With permission of White Arkitekter AB, Lund, Sweden.

A FREON-FREE ALTERNATIVE TO REFRIGERATION

Fig. 11.2 Earth cellar. With permission of White Arkitekter AB, Lund, Sweden.

serving food such as metal, plastic and paper are sorted for recycling. Urine and faeces are directed to separate treatment mechanisms where they are turned into liquid fertiliser and compost respectively. This fertiliser is sold to a local farmer who grows food for the school. Nutrients leaving the school are therefore eventually returned. Land has been set aside within the school for students and their families to grow food. Food grown within the school is stored in an on-site earth cellar (Fig. 11.2) that maintains a constantly cool year-round temperature without any mechanical cooling. Food from the school and local farms is then used in the school's kitchens.

The materials eco-cycle

The materials eco-cycle (Fig. 11.3) explains how the school has reduced its consumption of natural resources, particularly non-renewable materials through construction and operation. It also explains design consideration of the possible reuse of the building structure and materials in the future. The specification of construction materi-

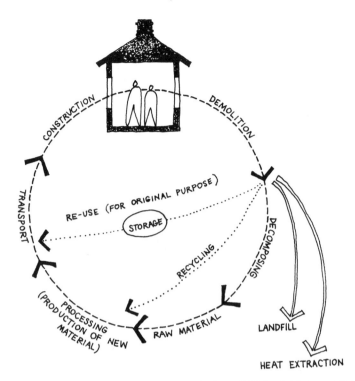

Fig. 11.3 Materials eco-cycle. With permission of White Arkitekter AB, Lund, Sweden.

als for the school minimised the use of non-renewable materials by incorporating salvaged materials from a nearby demolition project. The specification also minimised the use of compounds such as synthetic adhesives that are environmentally harmful and which reduce ease of deconstruction and salvaging of materials for future reuse.

The materials eco-cycle begins with the construction phase, the point at which the specified materials are ordered, transported and erected on-site. This process is repeated throughout the facility's life cycle through maintenance and refurbishment work. The next phase in the material eco-cycle is the demolition phase. At this point some materials can be salvaged for reuse, or sorted for recycling into new materials. The building design avoids the use of chemical compounds and composite construction techniques that reduce the ability for material reuse and recycling. Biodegradable material might become fertiliser for new organic materials, while unusable material is sent to landfill.

Reuse and recycling

The goal for Ostratornskolan was to minimise the use of raw materials through extensive use of secondhand and recycled materials. The designers extended this intention by designing the building so that materials could be easily salvaged at the end of its life. Measures taken include:

- use of 22 secondhand windows and 200 000 secondhand bricks;
- use of crushed concrete and brick as aggregate, base-course and drainage (90 m³ of crushed brick was used as topping for footpaths);
- designing few layers in the structure;
- using no composite materials;
- incorporating demountable structures (for example, none of the brick walls are load-bearing);
- use of durable, easy-to-maintain materials.

To ensure that building occupants and future contractors understand the intent of these measures, a 'black box' (Fig. 11.4) has been built into the entrance of one of the buildings. It contains information about the

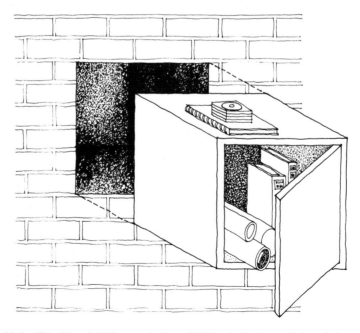

Fig. 11.4 'Black box'. With permission of White Arkitekter AB, Lund, Sweden.

products and materials that are used in the building, how they should be maintained, and measures that should be taken during future maintenance, construction and demolition to facilitate salvage.

Chemical products – long-lasting compounds

The design of the school buildings minimised the quantity of materials containing non-degradable compounds. A standard school specification in Sweden requires the use of about 80 tonnes of non-degradable compounds such as putty, floor adhesives, self levelling compounds, foam sealant and primer; enough material for two fully loaded trucks 24 metres in length. The school eliminated the use of PVC, CFC, HCFC and foam and reduced the amount of other non-degradable materials to about 4 tonnes, one-twentieth of the quantity found in a standard specification school.

The water eco-cycle

Ostratorn School uses natural processes, coupled with some clever engineering and simple technology to avoid water pollution and to provide nutrients for food production. The school's water eco-cycle (Fig. 11.5) works with nature's water cycle to reduce the school's use of mains supply, and eliminate connections to Lund's sewers. The results not only minimise disruption to natural water cycles, they reduce disruptions to nitrogen and phosphate cycles by removing nitrate and phosphate-rich human waste from wastewater and using it as fertiliser.

Rainwater as toilet water

The water eco-cycle has no real beginning or end, but at the Ostratorn School there is a convenient link between the top and our bottom. Ostratorn School's roofs collect rainwater which is then used to flush toilets. Rainwater is stored in two 9000-litre tanks in the basement. From here it is pumped via dedicated pipes to water-efficient split-pan toilets. The 18 000-litre holding capacity of the two tanks is enough water to flush water-efficient toilets for 300 people with an allowance built in for a two-month period of no rain. Had conventional flush toilets been used, the school would have required 324 000 litres of water to transport human waste to the sewage treatment plant.

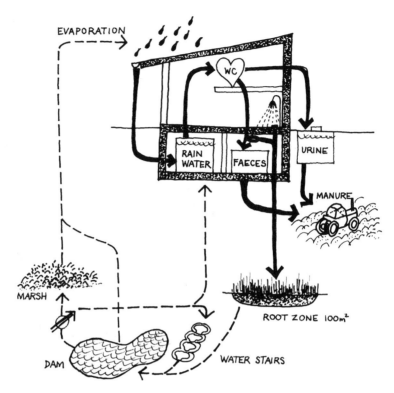

Fig. 11.5 Water eco-cycle. With permission of White Arkitekter AB, Lund, Sweden.

Turning human waste into fertiliser

The toilet pans are split to provide one receptacle for urine and another for faeces. The point of separating faeces and urine is to remove the nitrogen-rich urine from the wastewater stream where it causes pollution. Urine is flushed with 200 ml of rainwater, collected in one of the two 9000-litre underground tanks and, after a fermentation period, is sold to a local farmer as liquid fertiliser. Faeces are flushed with 4 litres of rainwater. A centrifuge device termed an 'aquatron' (Fig. 11.6) then separates water and faeces. Faeces is composted and sold as fertiliser to local farmers.

Using biology instead of technology

Grey water from basins and showers, stormwater from school grounds and the separated toilet water is all directed into a 100 m² reed-bed filtration area called the 'root zone' where aquatic plants biologically

Fig. 11.6 Aquatron, urine separator and composting toilet unit. Architect: White Arkitekter AB. Photograph by Tomas Lexén.

remove dangerous bacteria (Fig. 11.7). The water flows from the root zone through a cascading water staircase that oxygenates the water to continue the treatment process, and into a dam where it can be used to supplement rainwater for toilet flushing. All treated wastewater eventually ends up in a specially constructed wetland from where it evaporates and eventually falls again as rain.

The energy eco-cycle

Careful consideration of the major sources and uses of energy in the school's buildings was required to create the energy eco-cycle. The two major principles that became the building blocks for a design response were minimal energy consumption and maximum use of energy-quality. A school can heat up very easily because of the large number of people, equipment and lighting in the buildings. The main focus was therefore to efficiently remove excess heat and to use it where required for maintaining comfortable temperatures during winter. This was achieved using a heavy building frame, brick walls

Fig. 11.7 On-site grey-water treatment. (a) Root zone; (b) water staircase; (c) dam. Architect: White Arkitekter AB. Photographs by Tomas Lexén.

for thermal mass and large interior spaces to moderate day and night temperatures. Shading to the east and west facades helps reduce heat gain in spring and summer, while good natural light and low-energy light fittings minimise heat gain.

These considerations translated into four ecologically sustainable design strategies: natural ventilation, solar chimney, summer cooling and solar collector.

Natural ventilation

Ostratorn School's natural ventilation (Fig. 11.8) was required to:

- give fresh air without drafts;
- be adaptable to the number of people and activity in a space;
- consume no more energy than a mechanical system with heat recovery.

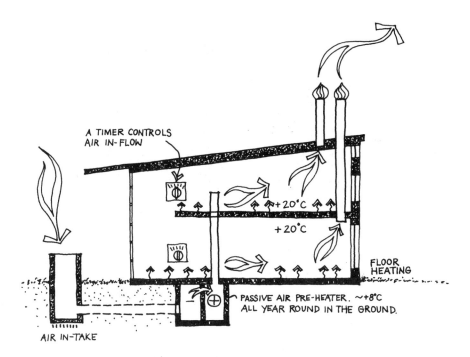

Fig. 11.8 Natural ventilation system. Diagram reproduced with permission of White Arkitekter AB, Lund, Sweden. Photograph by Tomas Lexén.

While some buildings are designed to maintain comfort despite their climate, this building is infused with it, using the properties of building materials and constant below-ground temperatures to maintain ventilation. Temperature differentials between outside air temperature, ground temperature and room temperature are harnessed to draw cool air through a 2 m high by 1 m wide subsurface concrete air intake shaft and into the buildings.

In winter, cold outside air is drawn into the subsurface shaft where temperature is a constant +8°C. This preheats incoming air (which can be as cold as –20°C) and reduces the energy required to heat air. Roof vents allow warm air to escape and help draw cool air into the building. Supplementary heating is provided by a district gas furnace.

In summer the subsurface air intake helps cool incoming air. However, on certain days in summer outside temperature can be the same as or higher than the inside temperature. Unless designed for, these conditions could draw cool air out of buildings rather than cooling the air that comes in. It is for this reason that Ostratorn School is equipped with a solar chimney.

The solar chimney

Natural ventilation requires a difference between inside and outside temperature. When outside temperature equals inside temperature, additional energy is required to achieve desired air changes. Ostratorn School has adopted an idea used in concert halls in Sweden in the nineteenth century to overcome the problem. When the outside and inside temperature differential was not adequate, fires were lit in the attics to draw air through the building. Ostratorn applies the same concept, replacing conventional chimneys with solar chimneys (Fig. 11.9). Directing solar hot water through pipes in ventilation stacks raises air temperature in the stacks to 45°C and creates the thermal differential required to drive the natural ventilation system.

Summer cooling

The floors of Ostratorn School do more than provide a platform for learning, they also act as a cooling device (Fig. 11.10). The school uses in-floor heating and cooling. Tubes in the concrete slabs carry warm air in winter and cool air in summer. For summer cooling the in-floor air tubes are connected to a fan-coil unit that works at night to extract air warmed by thermal transfer from people during the day. Cool

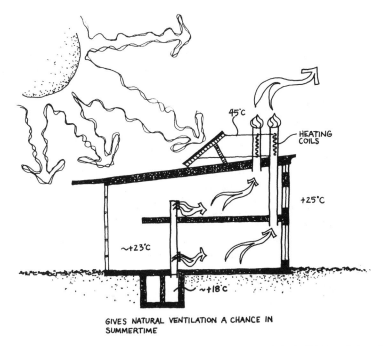

Fig. 11.9 Solar chimney. With permission of White Arkitekter AB, Lund, Sweden.

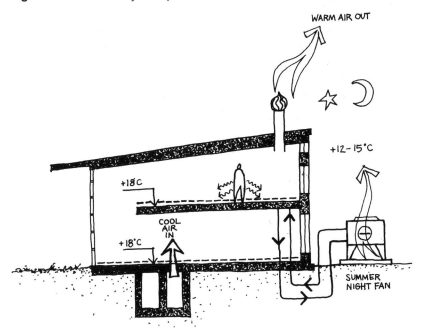

Fig. 11.10 Summer cooling system. With permission of White Arkitekter AB, Lund, Sweden.

night air lowers internal temperatures to 18°C by the time students and staff arrive the next morning.

Solar collector

Ostartorn School uses solar energy to provide its hot water. But instead of using roof-mounted solar collectors, the whole roof is designed as a solar collector (Fig. 11.11). This avoids maintenance and replacement costs and means that the roof structure is not only providing shelter. Water pipes are cast into about one-third of the surface area of the school's concrete roof. The solar hot water system is a closed circuit system. Water heated on the roof is passed through a heat exchanger in the basement to heat water for use. This approach keeps energy in the system longer and reduces the need to purchase energy for heating water.

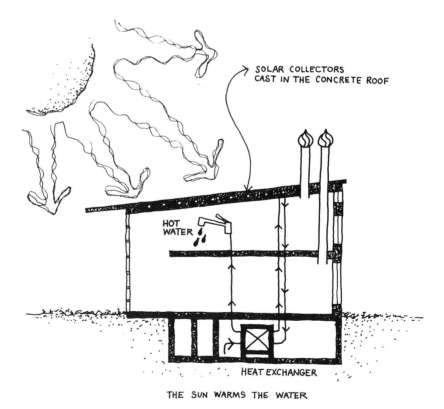

SOLAR COLLECTORS CAST IN THE CONCRETE ROOF

HOT WATER

HEAT EXCHANGER

THE SUN WARMS THE WATER

Fig. 11.11 Solar collector. With permission of White Arkitekter AB, Lund, Sweden.

Ecologically sustainable laws, principles and strategies

From Ostratorn School's design around eco-cycles, it is obvious that making the project ecologically sustainable was an important aim. But has the school achieved its goal? It is hard to determine the influence of one project on the ecosystems it affects, however we can determine whether the project has a design that accords with laws and principles of ecological sustainability. For this task we can use the laws and principles as a sustainability checklist.

Law: Create systems that consume maximum energy-quality
Use low-quality, highly abundant energy resources

The buildings' emphasis on maintaining a comfortable interior environment through passive strategies like providing thermal mass and designing for natural ventilation and natural light means that the buildings are using the low-quality energy of heat gradients and daylight to maintain comfort. Capturing highly abundant heat energy from the sun to heat water and air reduces high-quality, low-abundant fossil fuel consumption.

Even 'waste' heat from the interior of the building has been used. Heat radiated from lights, equipment and people is absorbed by thermal mass in floors, which in turn heats air in in-floor ducts. In summer the heat is drawn out of the building by a 'night fan' and transferred outside the building. In the winter this heat is retained to help keep the buildings warm.

Use energy in a large number of small steps, not in a small number of large steps

Implicit in the eco-cycles approach is consideration of the life cycle of energy in its many forms throughout the project. The material eco-cycle, for example, emphasises the use of secondhand materials, and considers materials in the buildings as resources for the future. Reusing and designing for the further reuse of materials in effect uses the initial embodied energy of materials again and again.

Energy is always added to building materials when they are re-used. At the school, the reuse of bricks has a double advantage of not only using secondhand material that requires little additional processing for reuse, but also providing a material with a high thermal mass. The additional processing energy is perhaps paid back over time by the ability of the bricks to store heat and contribute to passive design measures that maintain indoor comfort.

The secondhand windows, on the other hand, while saving initial embodied energy, are not as well insulated as the new ones used on the project. The architects are hoping that other energy-effective design features of the buildings will compensate for their inefficiency.[2]

In the energy eco-cycle the solar energy is used to heat water. The hot water is then used to heat incoming air in winter and outgoing air in summer through the solar chimney. This solar hot water also flows through a heat exchanger and is used to heat water for building occupants.

Minimise waste

Ostartorn School minimises waste in a number of ways. Firstly, by using secondhand materials it keeps materials in active service and diverts them from landfill sites. Secondly, by being designed from deconstruction and providing care instructions, the designers have provided the infrastructure to minimise waste of materials and the burden on landfill in the future.

Water is not wasted, either as it is consumed or as it is discarded. The use of water from reservoirs is reduced through water efficient fittings and supplementing rainwater for non-potable uses. This minimises the school's connection to any water wasted through leaks in supply infrastructure. The buildings treat and reuse the water that flows through the buildings, therefore removing the potential for the building to add polluted wastewater to the town's sewerage systems. High-grade water is used for drinking, low-grade water is used for toilet flushing and irrigation, and when it is not being used by humans it isn't being wasted either. It is providing habitat for insects, animals and plants in the dam and wetland.

Through the food eco-cycle, food waste and human excrement become fertiliser instead of polluted wastewater. Leftover packaging, food preparation, and serving materials are separated and recycled, diverting solid waste from landfill.

The waste of delivered energy is minimised through good insulation, including double glazing to all new windows. Wastage of dissipated energy is minimised by building with high thermal mass materials and using temperature gradients to drive natural ventilation.

Law: Consume resources no faster than the rate at which nature can replenish them

Maximise use of renewable and used resources

Solar energy is not used to generate electricity for the school. However, buildings are designed to maximise use of passive solar energy to minimise the need for fossil fuel-generated energy. Solar energy is used to heat water that is in turn used to heat water for occupant use and warm air. Incorporating water pipes in the roof slabs turned one-third of the roof into a solar collector, maximising exposure to sunlight.

In a similar way, rather than using wind to generate electricity, buildings are sited so that breezes help to cool buildings in summer. Trees and landscaping create microclimates that shield buildings from cold winter winds.

Constant soil temperatures might also be considered a renewable resource, or more accurately an 'undepletable' resource. The constant temperature of soil was used as a passive heat exchanger, pre-cooling air in summer, warming air in winter for free and reducing the demand for fossil fuel energy. Constant soil temperature is also used to store food in the earth cellar without refrigeration.

Water is a renewable resource if it remains unpolluted. Ostratorn School makes maximum use of the water it receives from Lund's reticulation systems and the water that falls as rain on the site by treating and reusing it on-site. This also removes the school's contribution to water pollution downstream of the school and therefore helps maintain good water supplies for the region.

Minimise use of new and non-renewable resources

Most of the materials used in construction are non-renewable. One possible strategy to adhere to this principle is therefore, to concentrate on lightweight, high-strength structural design that is as efficient as possible with materials chosen. The designers of Ostratorn School, on the other hand, have concentrated on mass rather than lightweight design, perhaps increasing demand for raw materials with the intention that their choice will decrease consumption of fossil fuel over the life of the building by providing thermal mass.

Another way the building could be seen to be 'paying back' initial material consumption is with the incorporation of secondhand bricks and windows, reducing the demand for new structural materials. Design for deconstruction also provides the opportunity to reduce demand for new materials in the future by making it easy to 'mine' the school's buildings for materials.

An emphasis of the design of the school is the minimisation of fossil fuel consumption. This is achieved by reducing demand for delivered energy through passive solar design, natural ventilation and the installation of energy-efficient light fittings and equipment. Water efficient fittings also contribute to reduced energy consumption by reducing the volume of water needing to be heated (remembering that solar hot water is supplemented by district heat generated using natural gas).

Perhaps a more subtle strategy to minimise fossil fuel consumption is the establishment of garden plots for students and their families to grow food in. If crops are successful the school's garden could contribute to reducing fuel energy consumed in producing, transporting, packaging, refrigerating and selling commercially produced food.

Do more with less

Doing more with less is the catchphrase of resource efficiency. The school does more with less in a number of ways. The roofs provide shelter, collect water for toilet flushing and act as solar hot water heaters, providing three services where an inefficient roof might only provide shelter. Light fittings, appliances and plumbing provide the required amenity while using far less energy and water than standard technology. Secondhand bricks provide structure and thermal mass with far less embodied energy than new bricks. The school's space and structure are designed to be able to be easily retrofitted should the buildings need to be used for something other than schooling, providing the possibility of more than one use from the expenditure of energy and materials required initially to build the school.

Law: Create only byproducts that are nutrients or raw materials for resource production

Eliminate pollution

Pollution is eliminated by:

- separating urine from faeces and turning both into fertiliser;
- removing phosphates and nitrates from wastewater where it would become pollution if allowed to flow into aquatic ecosystems;
- not using toxic materials or materials that create toxic pollution during manufacturing such as PVC, CFCs, HCFCs and foam;
- separating garbage for recycling instead of sending it to landfill where it becomes environmental pollution;
- providing garden plots to create the opportunity to grow organic food;
- biological treatment of wastewater for reuse on site to stop polluted water from the building ending up in natural watercourses.

Use biodegradable materials before bio-accumulating materials

Choosing materials and designs that required minimal finishing helped reduce the use of non-biodegradable compounds to one-twentieth of the level that is found in standard Swedish schools. Reducing quantities of metals proved harder. The reduced quantities of toxins in the buildings contribute to making materials easier and safer to reuse in the future.

The school makes use of its biodegradable waste by composting human excrement and food waste. When used as fertiliser by local farmers for food crops, this helps reduce the demand for commercially produced fertilisers, concentrations of which can accumulate in groundwater and aquatic environments.

Reuse, then refurbish and recycle materials, components and buildings

The eco-cycles approach to the design of Ostratorn School illustrates an intention to turn the outputs of consumption into inputs for production. Reuse of secondhand bricks and windows, and using crushed concrete as hardcore for paving, avoids creating waste, creates market demand for salvaged materials and helps provide incentives for the establishment of salvage and recycling services and infrastructure.

Design for deconstruction and the provision of a 'black box' containing material specifications and deconstruction instructions provides the opportunity for the building to be easily reused, refurbished or mined for its materials in the future. Water is reused and recycled throughout the buildings, as are the nutrients in the school's food through composting. Energy is reused as it degrades through heat exchangers, thermal mass, and natural ventilation.

Law: Enhance biological and functional adaptability and diversity
Apply life-cycle awareness and the precautionary principle

The eco-cycle approach shows the designers' consideration of life-cycle environmental performance issues at the school. The policy of 'making visible' environmental performance strategies is also important from a life-cycle perspective because it exposes young people to environmental innovation. There is no better place to teach the children of Lund about how to live sustainably with the Earth and to repair the damage done, than at their school. One of the best ways to facilitate this learning is by providing buildings that *are obviously different* to the buildings that are a normal part of a child's experience. If the school is warm without heating but their house isn't, the child can be prompted to ask 'why?' If the school uses rainwater to flush toilets and their home uses drinking water – again the question can be asked 'why?' If the school has panels on the roof to heat water or generate electricity – 'why?'

Provide access to fast-cycling materials without destroying slow-cycling materials

A number of strategies have been adopted to meet this criterion. These were:

- reducing the number of layers of material in the structure, using materials such as bricks, and dust-sealed concrete floors that required little finishing or additional components;
- eliminating mechanical ventilation and therefore extensive ductwork;
- minimising the use of adhesives, putties and sealants to keep materials 'clean' from hard-to-remove substances for recycling in the future;
- laying tiles in cement mortar so they can be easily removed without damaging the concrete slabs;
- the designers have provided instructions on deconstruction in order to ensure that these design features are not ignored and materials damaged or wasted in the future.

Protect and enhance biodiversity

Biodiversity is protected and enhanced both directly and indirectly as a result of this project. The use of the 'root zone', dam and wetland as biological water treatment processes directly provide new areas of habitat for plants and animals. Protection for remote ecosystems is offered through avoiding the use of toxic materials, composting wastes and recycling.

Summary

The approach taken to contributing to ecological sustainability is summarised in Table 11.1.

Table 11.1 Ostratorn case study summary.

Create systems that consume maximum energy-quality	
Principle no. 1	**Decision-making phase**
Use low-quality, highly abundant energy resources	
Strategies	
Create community gardens for local food production	Development planning
Natural ventilation	Design
Use high thermal mass	Design
In-floor slab air ducts for heat transfer	Design
Solar hot water	Feasibility
Use stable soil temperatures as a passive heat exchanger	Design
Principle no. 2	
Use energy in a large number of small steps, not in a small number of large steps	
Strategies	
Use heat exchangers to pre-heat or pre-cool incoming air and water	Design
Use secondhand materials	Design
Design for deconstruction	Design
High thermal mass	Design
Solar chimney	Design
Principle no. 3	
Minimise waste	
Strategies	
Reuse grey water	Development planning
Design for deconstruction	Design
'Black box' instructions for use and disassembly	Design
Composting organic waste	Operation
Solid waste recycling areas	Design/operation
New windows double-glazed and high thermal mass	Design

Table 11.1 (*Continued.*)

Consume resources no faster than the rate at which nature can replenish them	

Principle no. 1	**Decision-making phase**
Maximise use of renewable and used resources	
Strategies	
Use treated grey and black water for non-potable purposes	Development planning
Passive climatic design	Design
Use solar hot water	Design
Use materials and products that are easily recyclable	Design
Use secondhand or recycled materials	Design
Soil temperature used as passive heat exchanger	Design
Design for deconstruction	Design
Trees and landscaping provide sheltered microclimate	Design
Principle no. 2	
Minimise use of new and non-renewable resources	
Strategies	
Passive climatic design	Design
Choose low embodied energy materials	Design
Use secondhand or recycled materials	Design /construction
Design for deconstruction	Design
Use energy- and water-efficient appliances and fittings	Design
Community organic garden	Development planning
Principle no. 3	
Do more with less	
Strategies	
Incorporating solar hot water heating in roof structure	Design
Natural ventilation	Design
Use energy- and water-efficient appliances and fittings	Design
Use of secondhand bricks	Design
Few internal structural walls	Design

Table 11.1 (*Continued.*)

Create only byproducts that are nutrients or raw materials for resource production

---

Principle no. 1 Eliminate pollution	**Decision-making phase**
Strategies	
Separating urine and faeces for composting using split toilet pans	Design
Earth sheltered cool room	Development planning
Not using PVC, CFCs, HCFCs and foams	Design
Community organic garden	Development Planning
Biological wastewater treatment and recycling	Development planning
Principle no. 2 Use biodegradable materials before bio-accumulating materials	
Strategies	
Design and materials that require minimal finishing	Design
Not using PVC, CFCs, HCFCs and foams	Design
Composting organic waste for fertiliser	Development planning
Principle no. 3 Reuse, then refurbish and recycle materials, components and buildings	
Strategies	
Use of secondhand bricks and windows	Design
Design for deconstruction	Design
Use of crushed recycled concrete and bricks for paths and drainage	Design /construction /operation
'Black box' instructions for use and disassembly	Design
Biological wastewater treatment and recycling	Development planning
Use of heat exchangers and thermal mass to reuse heat energy	Design

Table 11.1 (*Continued.*)

Enhance biological and functional adaptability and diversity	
Principle no. 1 Apply life-cycle awareness and the precautionary principle	**Decision-making phase**
Strategies Using eco-cycles as a basis for design and operation Making environmental innovations visible to students	 Development planning Design
Principle no. 2 Provide access to fast-cycling materials without destroying slow-cycling materials	
Strategies Design for deconstruction Avoiding composite materials, especially for finishes	 Design Design
Principle no. 3 Protect and enhance biodiversity	
Strategies Community organic garden Biological wastewater treatment and recycling Not using PVC, CFCs, HCFCs and foams Solid waste separation and recycling	 Development planning Development planning Design Operation

Conclusion

The eco-cycles approach taken by BEEs involved in the Ostratorn School recognises the interdependency of building and nature, identifies ways of using thermodynamics to adhere to principles of ecologically sustainable building, and provides a life-cycle thinking model that considers the effects of change over time. The importance of dealing with ecological sustainability as early as possible in the building development process is reflected by the number of strategies incorporated into the project early in the development process.

Using ecologically sustainable building as the guiding approach to building development rather than incorporating it as an addition to a more 'traditional' non-ESD approach often leads to better results financially and ecologically.

More information

Laws and principles for ecological sustainability

> *The Natural Step for Business: wealth, ecology and the evolutionary corporation* (1999) Nattrass, B., Altomare, M. & Naijrass, B. New Society Publishers, Gabriola Island, Canada.
> *Ecological Design* (1996) Van der Ryn, S. & Cowan, S. Island Press, Washington DC, USA.
> *Designing With Nature: the ecological basis for architectural design* (1995) Yeang, K. McGraw Hill, New York.

Urban sustainability

> *Urban Future 21: A Global Agenda for 21st Century Cities* (2000) Hall, P. & Pfeiffer, U. E. & F.N. Spon, London, UK.
> *Ecocity Theory: conceiving the foundations.* In: *Village wisdom future cities – proceedings of the third international ecocity and ecovillage conference* Register, R. & Peeks, B. (1997). 8–12 January 1996, Yoff, Senegal. Ecocity Builders, Oakland, USA.

Case studies

> *Green Buildings Pay* (1998) Edwards, B. (ed) E. & F.N. Spon, London, UK.
> *Green Development: integrating ecology and real estate* (1998) Wilson, A., Uncapher, J., McManigal, L., Lovins, L.H. & Hunter, L. John Wiley, New York.
> *BDP Environment Design Guide* Royal Australian Institute of Architects Melbourne, Australia.
> *The Green Building Challenge* (1998, 2000) Over 30 international case-studies available on CD-Rom from www.greenbuilding.ca.

References

1 White Arkitekter (1998) *Ostratornskolan, Lund Sweden – A school built on ecological ideas.* White Arkitekter AB, Lund Sweden.
2 White Arkitekter (1999) – Interview October 2001.

12 FROM KNOWLEDGE TO UNDERSTANDING

Introduction

The principal aim of this book has been to learn what BEEs know about ecologically sustainable building. To know something is, however, not the same as understanding it. Knowing, for example, that the energy consumption of existing buildings in Australia are responsible for generating about 80.9 Mt of carbon dioxide equivalent greenhouse gas emissions per year,[1] is as abstract as knowing that certain species of ants in the Amazon rainforest grow fungus as food on caterpillar droppings while others grow food on decaying leaves[2] – unless we have a way to understand what the knowledge means.

Building ecology provides this means of understanding by relating our knowledge of the environment and our knowledge of building, with validated knowledge concerning ecological sustainability. Our knowledge of building needs to be related with, for example, the knowledge that one of the outcomes of species (ants, for example) using low-quality, highly abundant energy sources (like leaves) is the long-term accumulation of higher-quality resources (natural capital), while use of high-quality energy sources (like caterpillar poo) leads to depletion of natural capital and limited capacity to support large populations.[3,4]

If we, as BEEs, intend to create ecologically sustainable buildings then we must analyse these specific types of knowledge with an intention to find out what each can teach us about ecological sustainability. We can observe for example, that we, like the poo-farming ants, use high-quality, low-abundance energy sources (in our case, fuels like coal and oil) to run our buildings and thus contribute to depletion of natural resources, and ultimately to the inability to support increasing consumption of these energy sources. Unlike the leaf-farming ants, our energy sources are not abundant and renewable. Therefore, every new building connected to fossil fuel energy contributes further to our predicament. With this understanding, headlines proclaiming 'building boom' would be regarded more with a courageous determination to change current practice, than with an ecstatic reception for the promise of potential profits.

Ecologically sustainable buildings are a product of the relationships created by people making decisions as they transform an idea for a building into a physical structure. And how we integrate our understanding of building ecology in our own minds is most important in creating lasting change. Being able to understand the potential influence of decisions on both built and natural systems is a way of thinking that directs action. Our ability to think and then act in an ecologically sustainable way is indicator of our ecological literacy. Ecological literacy allows us not only the ability to identify problems with current practice, but also to devise solutions and lead innovation in sustainable building.

Throughout this book we have endeavoured to improve our ecological literacy and environmental awareness, to learn what BEEs understand about ecologically sustainable building. The principal shift required in our thinking is to move from an object-focused understanding of building to a systems perspective. In order to do this we set about understanding:

- the interdependencies of building and nature;
- how building affects nature;
- what is and is not ecologically sustainable;
- how building can work with nature.

Synthesising these understandings allows us to see the environmental implications of our decisions. So, what can our understanding of the way nature works tell us about how we can shape an ecologically sustainable built environment? How can our understanding lead us to build within the limits of the carrying capacities of ecosystems, the natural cycles that sustain them and the natural laws that govern them? Let's reflect on what we have learned.

Why building affects nature

First of all, building is interdependent with nature. We need to take a life-cycle perspective when we make decisions. We need to consider every material's life-cycle 'story'. This means understanding the environmental implications of materials ingredients, and the waste and pollution associated with mining. We also need to consider how far the ingredients have travelled, and how much energy has been used in processing them and manufacturing the material. We need to know how the material performs in use, how much maintenance is required, what materials are required to maintain it, how easily it can be reused or recycled, and whether it biodegrades or bio-accumulates when discarded.

Our choice of development land and the infrastructure for supplying energy, water, transportation and food all affect nature. We rely on the natural environment to assimilate the byproducts of our building construction, operation, alteration and demolition. An awareness of these interdependencies needs to be brought to bear on decisions we make about the way we consume resources and how we process wastes. We can, for example, choose to limit our connection to resource supply systems that are inefficient with precious resources like water, and our connection to waste infrastructure that causes pollution rather than encourages reuse. Finally, BEEs understand that building is an activity that relies on ecosystems for materials and services and that buildings and built environments are parts of ecosystems, not separate from nature.

How building affects nature

With an awareness of interdependency we can understand why building impacts on the environment. To understand *how* building causes impacts we turn to our understanding of natural cycles. We can see, for example, that vast quantities of energy have been required for building material production and supply, building construction and building operation. This energy has been supplied predominantly by burning fossil fuels, either to produce electricity or to run heavy machinery. The resulting greenhouse gas emissions are increasing in atmospheric concentration and contributing to global climate change.

The water cycle is changed by the dynamics of urban environments. Paved surfaces increase surface run-off, stormwater drainage channels water and pollutants from streets to natural watercourses, water from catchments often travels hundreds of kilometres only to be used once to flush a toilet. Water contaminated by pathogens and toxins is then transferred via pipelines to treatment plants or directly to outfall pipes, often discharging into coastal ecosystems. The nitrogen and phosphate cycles are altered by nutrient-rich urban and agricultural run-off and sewage discharge. The atmospheric phase of the nitrogen cycle is also affected by building, with the manufacture of materials like cement and steel causing the release of nitrous oxides. The manufacture of building materials like copper and the burning of fossil fuels to provide energy for building also affect the sulphur cycle. Increased concentrations of sulphur oxides in the atmosphere have mixed with water molecules, and fall as acid rain, damaging crops and buildings.

Ecosystems have been impacted by these changes in biogeochemical cycles. At the same time biodiversity, the quality of ecosystems that provides resilience to change, is being lost. Perhaps the greatest influence of building on biodiversity is the use of rainforest and old-growth timbers. Old-growth and tropical rainforest are among the most mature and complex ecosystems on the planet, yet rainforest species of timber such as meranti, kapur and western red cedar are commonly used in building.

With knowledge of natural cycles, ecosystems and the effects of building on them, we can develop a more comprehensive understanding of the life-cycle environmental effects of materials and buildings and can therefore choose materials wisely to limit our disruptive effect on natural systems. We can also change the way building is designed so that the environments we create don't help channel substances into natural environments at rates or in quantities that can't be assimilated by natural cycles.

What makes building ecologically sustainable?

Knowing why and how building affects the environment is interesting but, without a unifying objective that establishes what to do with this knowledge to improve the world, it remains untapped potential. The task of sustaining healthy ecosystems is fundamental to maintaining the biological basis from which we live our lives. Building in ways that sustain ecosystems is therefore the unifying fundamental objective. Thermodynamics and an understanding of the dynamics of change explain how building can be ecologically sustainable.

The first law of thermodynamics teaches us that energy can't be created or destroyed. This law focuses attention on how efficiently we use our materials and energy. As we cannot currently create more of the finite resources that we require for building, we need to ensure that materials and energy are not being wasted, that we consume as little material and energy to solve as many problems as possible, that we are doing more with less. The first law also focuses attention on the nature of the substances we use in building. If we use material that can't be transformed into harmless substances by nature once they have left our systems, it is obvious that harmful substances will build up in the environment. It reminds us that there is no such place as 'away'. Similarly, if we use substances that cannot be recreated as quickly as they are used, they will become less abundant in the environment. The first law instructs us that in order to sustain healthy environments and access to life-supporting resources, we need to use material and energy resources that do not pollute and that are easy to regenerate.

The second law of thermodynamics teaches us that as energy moves and is transformed through a system its ability to do work (its quality) decreases. Because of the first law, the total amount of energy remains constant, but the amount of low-quality energy increases, tending to decrease order in the system, a phenomenon described as increasing entropy. The imperative is to keep useful energy in the system for as long as possible and to devise ways of using energy at each stage of its descent down the quality scale. The second law therefore challenges us to consider the quality of energy we choose to use and to make the best use of it. While the first law emphasises energy efficiency, the second law emphasises energy effectiveness.

To increase order in a system, high-quality energy needs to be added. If you consider Earth to be the total system, it is easy to see that the additional energy that maintains order comes as low-quality abundant energy from the sun. Our built environment is another system requiring energy to maintain order. Our energy comes mainly from finite stores of very high quality energy in the natural environment. Therefore, as we import energy to maintain or increase order in our built environment, we decrease order in natural environments. The second law also shows us that we can help maintain order in natural systems by shifting from finite high-quality energy to low-quality, highly abundant energy.

Natural systems achieve both resource efficiency and resource effectiveness by ensuring that energy of different qualities is kept in the system for as long as it can be used. The mechanism for ecologically sustainable energy and resource use, according to the fourth law of thermodynamics, is the keeping of energy quality in systems through the development of feedback loops. Sustainable natural systems have evolved ways of turning the outputs of consumption into resources for production, thus reducing the need for the system's reliance on energy from outside sources. In other words, ecologically sustainable systems use feedback loops to turn waste from consumption into food for production. Positive feedback is the key to surviving designs.

So from a thermodynamic perspective an ecologically sustainable building:

- consumes resources no faster than the rate at which nature can replenish them;
- consumes maximum energy-quality;
- creates byproducts that are nutrients or raw materials for resource production.

Buildings that adhere to these laws are designed to do more with less, not to use polluting materials, to use resources effectively by consuming as much energy-quality as possible, to use renewable resources in preference to non-renewable. They are also designed to keep energy and resources in the system for as long as possible using feedback loops.

Ecologically sustainable building must also have the ability to adapt to change and to protect the qualities of ecosystems that allow them to deal with change. Sustainability, as the Oxford dictionary defines it, means to maintain or prolong. In the context of building we need to be careful not to interpret this as needing to resist change. One way of thinking about an ecologically sustainable building is that, rather than being something that lasts forever, it is something that maintains its amenity while adapting to change. Ecologically sustainable building must avoid obsolescence by being adaptable to the changing expectations of building users and the changing pressures of the natural environment. To be adaptable to change, buildings must be flexible in use and easy to alter. They must also be designed for their eventual demise, allowing disassembly of components and materials for reuse or recycling.

To be effective in an ever-changing environment, ecologically sustainable building must:

enhance biological and functional adaptability and diversity.

How to build with nature

How do we build with nature? What have we learned? We have learned that we need to think differently about building – regarding a building as the outcome of relationships we create rather than the structures we construct. This shifts our perception from objects to systems and allows us to understand the implications our decisions hold for nature. We also see the implications that change in natural systems holds for us. We understand that building is part of nature, and that buildings are a part of ecosystems. We can see the building in this piece of paper.

We begin always by thinking about the effects of our decisions on whole systems, rather than the parts. We ask questions about the life-cycle implications of building and building material, considering long-term as well as immediate environmental implications. We make decisions that assist ecosystems in providing life-supporting goods and services. We build with natural laws and we take advantage of the renewable resources available to us.

We use every building development as a learning opportunity, not just for ourselves, but also for our project partners and our communities. We take action to change destructive practices and we use our new ecological understanding to develop new practices. We get involved in the movement for sustainable construction. We conceive, nurture, promote and facilitate the kinds of changes required in the industry to make building ecologically sustainable. We help others do the same.

References

1 Commonwealth of Australia (1999) *Australian commercial building sector greenhouse gas emissions 1990–1910, executive summary report.* Australian Greenhouse Office, Canberra. p.13.
2 Allen, T. (2002) Applying the principles of ecological emergence to building design and construction. In: Kibert, J., Sendzimir, J. & Guy, B. (eds) *Construction Ecology: Nature as the basis for green buildings.* Spon Press, New York. p.115.
3 Odum, H.T. (1996) *Environmental accounting: EMERGY and environmental decision making.* John Wiley & Sons, New York.
4 Allen, T. (2002) Applying the principles of ecological emergence to building design and construction. In: Kibert, J., Sendzimir, J. & Guy, B. (eds) *Construction Ecology: Nature as the basis for green buildings.* Spon Press, New York.

GLOSSARIES

Glossary of terms

Agenda 21   A strategy developed at the Rio Earth Summit in 1992 that sets out a strategy for global action on sustainable development with an emphasis on human interactions with environmental problems. It emphasises the importance of public participation in environmental decision-making.[1]

Appropriate technology   A technology that responds to the needs and capabilities of the community using it, and which respects natural systems and local social and cultural practices.

Biodiversity   The diversity of genetic material, species and ecosystems in a given area.

Biogeochemical cycles   The mechanisms which circulate matter through the Earth's biosphere and atmosphere.

Bioregion   A geographical area defined by a group of co-dependent ecosystems.

Biosphere   The 'layer' of the air, water and land in which life exists or is supported.

Brownfield   Descriptive of a building site that has been contaminated to some extent.

Building ecology   The study of the interdependencies of building and nature.

Building professionals   People involved in any stage of the building development life cycle. Property developers, quantity surveyors, architects, engineers, project managers, builders and tradespeople are all considered building professionals.

Carrying capacity   The ability of an ecosystem to assimilate waste and pollution, or cope with habitat damage. The threshold at which any additional human activity would cause ecological damage. 'The maximum number of any species that can be supported by an ecosystem on a long-term basis.'[2]

273

CO$_2$ equivalent (CO$_2$$^{-e}$)   Carbon dioxide is used as a common unit to quantify greenhouse gas emissions. The CO$_2$$^{-e}$ measure includes all CO$_2$ emissions released during the production, processing and transportation of an energy resource, and includes an allowance for the methane and nitrous oxides released.[3]

Co-generation   The process of generating power and heat from the same process. Large facilities such as hospitals often have co-generation plant where gas is burnt to run electricity generators and the heat produced during combustion of the fuel is used to heat water for other services such as space heating.

Delivered energy   Energy that is transmitted for consumption, such as electricity or petroleum.

Design guidelines   Published guides to environmental design including data sheets on methods and issues, green building case studies, and environmentally friendly product databases.

Dissipative structures   The systems of relationships in ecosystems that are created to use energy-quality.

Ecological literacy   Ecological literacy relates with the concept of environmental literacy. As Strauss explains: 'An environmentally literate person recognises that human actions have complex ecological and normative consequences. He or she has the motivation and education to investigate and pursue courses of action that contribute to a more sustainable future.'[4]

Ecological rucksack   The amount of waste material produced during mining and processing of mineral resources as a proportion of the quantity of finished product produced.

Ecologically productive   Descriptive of land that can support or is supporting habitat.

Ecologically sustainable development (ESD)   There are many definitions of ESD. The Australian National Strategy for ESD defines it as: 'Using, conserving and enhancing the community's resources so that ecological processes, on which life depends, are maintained and quality of life for both present and future generations is increased'.[5]

Ecology   The study of the relationships between organisms and the environment. [from the Greek *oikos* 'house' and *logia* 'study']

Ecosphere   The layer of the biosphere that contains plants, animals and other life forms – both dead and alive.

Ecosystems   Systems of biological communities in their physical environments. Ecosystems provide essential life support goods and services such as clean air, fresh water, food and waste assimilation.

Ecosystems resilience   The resilience of an ecosystem refers to the system's ability to maintain functioning under changing circumstances. Ecosystems are considered resilient 'when ecological interactions reinforce one another and dampen disruption'.[6]

Embodied energy   The amount of energy required to create a product. As Tucker *et al.* define it, embodied energy includes all energy consumed from all sources for '…mining, transporting and processing of raw materials to the final delivery of the product, including the energy of all intermediate transporting and manufacturing processes and the share of energy required to provide the capital infrastructure which enabled the product to be produced'.[7]

Energy modelling   Predicting the energy demand of building designs for the purpose of improving energy efficiency. Energy modelling is normally conducted using sophisticated computer programs. These programs normally concentrate on providing information on life-cycle operational energy-related performance of buildings.

Energy-quality   Also referred to as exergy, energy grade or free energy. The ability of an energy source to do work. Our engines, turbines, cars and buildings consume energy-quality, degrading energy from high quality to low quality. The lower the energy-quality, the harder it is to use that energy source to do work.

Entropy   A measure of disorder in a physical system.

Environmental performance   Ultimately the environmental performance of a building is a function of how well it minimises all negative environmental impacts, the benefit to the community of the process of its development and its operation, and the level of amenity, comfort and well-being it provides for its users. Environmental performance may also be measured against the environmental goals of a project as set by the project stakeholders. Environmental performance requirements may therefore be more limited in scope than the above definition. The environmental performance of a building is often assessed against criteria such as energy efficiency, greenhouse gas emissions, resource consumption and waste production.

Environmentally aware   Environmental awareness relates with environmental literacy – see **ecological literacy**.

Exergy   See **energy-quality**.

Extended metabolism approach   A methodology for assessing the resource flows through an urban area as well as the effects of urban infrastructure on resource flows. It enables an assessment of the environmental impacts associated with the patterns of resource use to be identified.

Gaia hypothesis The Gaia hypothesis, initiated by Lovelock (1972)[8] and then developed by Lovelock and Margulis (1974),[9] asserts that the Earth is a self-regulating entity in which living and non-living parts are interlinked by complex feedback loops. The self-organising effect of these interacting elements maintains conditions for life to exist. It takes its name from the Greek earth goddess Gaia.

Greenhouse gases Gases such as carbon dioxide, nitrous oxides and methane, which absorb radiant heat from earth and therefore help to keep Earth's surface temperature within a range that supports life. Rapid increases in the atmospheric concentration of greenhouse gases (due to increased emissions from human industrial activity) is increasing the absorption of radiant heat, which in turn is being experienced as 'global warming'.

Grey water Water that has been used for domestic purposes such as clothes washing and cleaning. Grey water often contains detergent residues and has high concentrations of phosphates. Grey water can be used without treatment for irrigating non-food gardens and can be reticulated after treatment for reuse in toilets and for general washing.

Heat-island effect When radiant solar radiation from urban areas causes locally higher air temperatures.

HVAC An acronym for mechanical heating, ventilation and cooling systems in buildings.

Informal settlement An unplanned settlement, consisting of poor-quality or shanty houses and little if any basic infrastructure. Many informal settlements develop around cities in developing countries. People in these settlements usually have no security of land tenure.

Kyoto Protocol The Kyoto Protocol is an addendum to the United Nations Framework Convention on Climate Change (UNFCCC) adopted in 1997 that sets out stringent measures to be taken by developed countries to reduce their greenhouse gas emissions by an agreed amount relative to emission levels in 1990. The Protocol is yet to be ratified. At the date of this publication Japan, the European Union and the United Kingdom had agreed to sign up to the Protocol. The United States, and Australia (who negotiated an increase in emission levels under the Protocol of 8%), have refused to sign on the grounds that the conditions of the Protocol would impede economic growth in their countries. Some developed countries also complain that the Kyoto Protocol should also apply to developing countries.

Material supply chain   All of the processes and organisations involved in creating a building material and delivering it for use.

Multidisciplinary   Where there are people with different professional abilities.

Multi-stakeholder   Where there are a number of people with an interest in, or who are affected by, a project.

Normalisation   Normalisation is 'the process of factoring results by a common unit (persons, years, area, and so on) in order to allow for a fair comparison of the relative significance of impacts. Normalisation is sometimes required in preparation for additional procedures such as weighting or interpretation.'[10]

Operational energy   The energy required to run a building once it has been constructed and is in use.

Primary energy   Energy consumed to produce deliverable energy such as electricity or petroleum.

Self-organisation   The spontaneous emergence of order in systems. A constant flow of energy and matter through the system is necessary for self-organisation to occur.

Sewage   Sewage is the name given to the effluent that flows through sewerage systems, hopefully to a treatment plant.

Sick building syndrome   A condition marked by headaches, respiratory problems, etc. affecting office workers, attributed to factors such as poor ventilation in the working environment.

Stakeholders   People or organisations affected by a building development

Systems perspective   A systems perspective considers the relationships between elements rather than focusing solely on the elements. It recognises the influence of patterns of organisation on how the world works, not just the properties of substances.

Thermal mass   The potential capacity of a building assembly or system to store heat. Adobe walls and concrete floors have a high thermal mass while windows and aluminium cladding have a low thermal mass.

Trophic levels   The stages in the movement of energy through an ecosystem. Trophic levels also describe an organism's 'feeding status' in an ecosystem.[11]

Waste water   Water that is considered to have no further useful purpose.

Watershed   The water catchment area of a given location.

Glossary of selected strategies

This glossary explains selected strategies listed in the text. They are arranged alphabetically and the phase in the building development process when they are normally implemented is noted.

Bio-climatic design (Development phase: design)   Using building elements and the characteristics of building occupants to modify the indoor climate without special equipment by controlling access of heat and light from the sun,[12] using local climatic conditions such as prevailing winds, site topography, etc.

Bioregional planning (Development phase: planning)   Bioregional planning integrates urban and natural environments and promotes the design of communities to function in harmony with the ecosystems within which they are geographically located. Bioregional planning emphasises self-reliance, the use of indigenous landscapes, waste elimination through recycling and co-generation, and encouraging local economies and employment.

Building environmental performance rating schemes (Development phase: design and operation)   Assessment programmes that rate the environmental performance of a completed building or building design. Some rating schemes concentrate on single environmental performance criteria such as greenhouse gas emissions (for example, see Australia's Sustainable Energy Development Authority Greenhouse Rating Scheme www.seda.nsw.gov.au). These schemes can also be used as a design guide.

The Building Research Establishment in Britain was the innovator of multiple criteria rating schemes, creating the British Research Establishment Environmental Assessment Model (BREEAM) (see www.bre.com.uk). The USA has also developed a multi-criteria building rating scheme known as LEED (Leadership in Energy and Environmental Design) (see www.usgbc.org). A more technical system called the Environmental Index is offered by the Dutch Institute for Building Biology and Ecology (NIBE) (see www.NIBE.org). Some have criticised these schemes for focusing too heavily on developed world issues and for not incorporating an assessment of the social aspects of sustainable building. The Sustainable Building Assessment Tool (SBAT) assessment scheme developed by CSIR in South Africa adds an assessment of a building's social as well as environmental performance (see www.csir.org.za) for contact information. There are many rating schemes; each has its limitations and is only relevant to the country for which it has been developed.

Design for deconstruction (Development phase: design) Buildings designed for deconstruction can be dismantled, and the materials and components easily separated for reuse or recycling. Design for deconstruction can be, and is in many cases, also applied to temporary structures required during construction and to elements such as formwork systems.

Some of the methods used to achieve deconstructability include the use of bolted connections instead of ridged welded or cast-in connections. The use of prefabricated elements also adds to the ease with which a building can be dismantled. Consideration must also be given to the durability of materials. These methods potentially reduce the time and cost of refurbishment or demolition and help to maximise the reuse of resources.

Deconstruction instructions (Development phase: design and construction) Instructions on how to best remove building materials or building elements for reuse.

Eco-city planning (Development phase: planning) Eco-city planning recognises that sustainable urban systems are a dynamic interrelationship between economic, social and ecological factors. Eco-cities are designed and built based on ESD principles and foster community involvement and promoting an awareness of ecological issues. Developers of eco-cities consider issues such as the ecological impact of the development, reducing the need for the use of cars, emphasising the use of public transport and bicycles and designing amenities which are easily accessible by foot. Repairing and protecting urban ecology and promoting local food production is also important. Eco-city builders consider the sustainability of natural and urban environments when they design buildings. They consider the effect of each building on social, economic and ecological systems and consider how buildings can contribute to sustainability rather than simply how they can minimise environmental damage.

Eco-design (Development phase: design) Also Ecological design – Designing in accordance with ESD principles.

Ecologically sustainable material purchasing policies (Development phase: design, construction, operation and disposal) A policy to purchase only 'low impact materials'. These are materials that have the least negative impact on people and ecosystems, while still meeting functional and economic requirements. In essence, the best material or product for the job is the cheapest in ecological and social costs in addition to it being the most financially cost-effective.

More sustainable materials are those which come from a sustainably managed source, are non- or low-toxic, cause minimal ecological impacts, are low in embodied energy, are durable, and which require little maintenance. BEEs must find these qualities in materials that meet performance standards, as well as the aesthetic and economic expectations of the client. Databases listing environmentally friendly materials are available in many countries. Some examples include:

- Australia: 'Eco-Specifier' http://ecospecifier.rmit.edu.au
- Great Britain: Environmental Profiles
  http://collaborate.bre.co.uk/envprofiles/
- International: Forest Stewardship Council (sourcing timber from sustainably managed forests) http://www.fscoax.org/

Environmental cost–benefit analysis (Development phase: feasibility)   Cost–benefit analysis has been used to determine the financial and social costs and benefits of developments like a proposed third airport for London[13] and the Alcoa aluminium smelter at Portland in Victoria.[14] In all cases social and environmental costs and benefits are expressed in dollars. The basis for valuing the cost of resources used in a development is therefore the value of 'social sacrifices' in not having them available for alternative uses.[15] Recent research contends that that the cost of environmental impacts could be brought to bear on the results of a cost–benefit analysis through weighting of issues and the use of a zero or compounding rate rather than a discounting rate to account for higher values placed on environmental resources in the future, and the effects of environmental impacts being increasingly significant as time goes on. This new rate has been called the 'sustainability constraint'.[16]

Environmental impact assessment (Development phase: feasibility)   Environmental impact assessment is a social process developed to examine the potential environmental impacts of a project. Its main function is to determine environmental risks posed by a development alternative and to decide whether to proceed with the development and under what conditions.[17]

Environmental management (Development phase: construction, operation and disposal)   Environmental management is a formalised approach to controlling the environmental impacts of an organisation, whether the impacts are generated within the organisation, or are associated with the interaction of the organisation's activities and the external environment. It is an organisation's formal structure that includes a policy regarding the environment, as well as the procedures, resources, and those responsible for the implementation of environmental management.[18]

Environmental management planning is a process for identifying the environmental impacts of an organisation, formulating goals, objectives and procedures, and implementing a plan to reduce those environmental impacts. Implementation is a process of continual improvement where the effectiveness of the plan is monitored and results feed back to management for both internal and external auditing. The development and operation of environmental management systems is subject to the ISO 14000 series standards.

Expose resource supply and disposal pathways (Development phase: design)   We normally hide the water and gas mains, the stormwater and sewerage pipes. Exposing parts of these services, so that the links between source or resources and sink for wastes becomes more obvious, helps promote systems thinking.

Habitat and species surveys (Development phase: feasibility)   Processes for identifying areas inhabited by wildlife. These surveys are often carried out by government agencies charged with environmental protection. Particular concern is given to identifying rare habitat and endangered species so that their protection can be incorporated into building development plans.

Life-cycle assessment (Development phase: design)   A method for compilation evaluation of the inputs, outputs and the potential environmental impacts of a product system throughout its life cycle.[19]

Life-cycle cost planning (Development phase: feasibility and design)   Also known as life-cost planning,[20] is a method for evaluating both the capital and recurring costs of a project at the feasibility and design phase. Traditional cost planning focuses on all costs associated with a building project up to the stage of handover. Including calculation of recurring costs at the design phase allows the project team to factor in financial savings associated with providing higher environmental performance. When measures taken to improve the environmental performance of a building increase its capital cost, life-cost planning can show any associated financial savings and determine a 'pay-back' period. The pay-back period is the amount of time it takes for savings in building operating costs to equal any additional capital costs associated with environmental performance.

Long-life loose fit design (Development phase: design)   This life-cycle design philosophy requires designers to think beyond the client's initial requirements and envisage not only the construction and economic life cycles of a building, but also consideration of possible future use, periodic maintenance, refurbishment, and eventual demolition.

One key aspect of this strategy is designing to allow flexibility in use. Designing durable structures and flexible interiors can allow, for example, offices to become apartments, shop houses to become restaurants, or detached housing to become medical or commercial office space. This minimises resource consumption by avoiding demolition, saves energy by not having to reconstruct the structure, and in many cases helps preserve culturally significant buildings.

Life-cycle design can also be applied to housing design in order to respond to changing family needs. Houses can be designed to allow cost-effective alteration or expansion to cater for first home buyers, growing or extended families, and the elderly, without the need to build a different house for each situation.

Operating and maintenance instructions for the building as a system (Development phase: operation) Instructions for how the building should be operated as a system in order to maximise environmental benefits. Instructions should cover the use of building systems like ventilation, cooling and heating systems, as well as how to use the structure of the building and the properties of the materials that it contains. Instructions might include how to operate passive ventilation systems by opening and closing windows or vents in different areas to enhance natural ventilation.

Passive solar design (Development phase: design) See **bio-climatic design**.

Refurbish rather than build new (Development phase: feasibility) Modifying existing buildings for reuse, rather than assuming a client's needs can only be met through the construction of a new building.

Waste minimisation planning (Development phase: construction, operation and disposal) Waste minimisation planning is a systematic approach to ensuring that a building produces minimal waste. The planning process involves: identifying possible causes of waste and waste streams by conducting an audit; choosing appropriate design, procurement and management responses to minimise waste; setting challenging and achievable goals for employees to encourage waste minimising behaviour; monitoring performance; providing feedback; and recognising good performance. The best results for the environment and financially are achieved when waste is avoided. The waste minimisation approach is therefore to avoid (or refuse) waste, then reduce, reuse and recycle.

Other glossaries

International Energy Agency Annex 31

http://annex31.wiwi.uni-karlsruhe.de/ANNEX_XXXI/glossary.htm
Provides a comprehensive glossary of terms for environmental performance assessment of buildings. It was developed by an international team of researchers representing fourteen countries. The glossary is downloadable and is written in most European languages and Japanese.

BDP Environment Design Guide (Australia)

http://www.raia.com.au
A useful guide for architects and building designers.

References

1 Harding, R. (ed.) (1998) *Environmental Decision-Making: the roles of scientists, engineers and the public.* The Federation Press, Leichhardt, Australia. p.110.

2 Cunningham, P. & Saigo, B. (1997) *Environmental Science: a global concern.* Fourth Edition. McGraw Hill, New York. p.612.

3 Australian Greenhouse Office (1999) *Scoping study of minimum energy performance requirements for incorporation into the building code of Australia.* Australian Greenhouse Office, Canberra, Australia. p.75.

4 Strauss, B.H. (1996) *The Class of 2000 Report: environmental education, practice and activism on campus.* Nathan Cummings Foundation, New York. p.8.

5 Commonwealth of Australia (1993) *The National Strategy for Ecologically Sustainable Development (NSESD).* Canberra, Australia.

6 Peterson, G., Allen, C. & Holling, C. (1998) Ecological Resilience, Biodiversity, and Scale. *Ecosystems* Vol. 1. Springer-Verlag. p.11.

7 Tucker S.N., Salomonsson, G., Treloar, G., MacSporran, C. & Flood, J. (1993) *The Environmental Impact of Energy Embodied in Construction.* Main Report for the Research Institute of Innovative Technology for the Earth. CSIRO Division of Building, Construction and Engineering. Highett, Australia. p.21.

8 Lovelock, J. (1972) Gaia as seen through the atmosphere. *Atmospheric Environment* **6** 579.

9 Lovelock, J. & Magulis, L. (1974) Biological modulation of the Earth's atmosphere. *Icarus* Vol. 21.

10 International Energy Agency – Annex 31 (2002) *A Glossary of Key Terms in English, French, German, and Japanese.* p.6. Downloaded on 11 June 2002 from: http://annex31.wiwi.uni-karlsruhe.de/ANNEX_XXXI/glossary.htm.

11 Cunningham, P. & Saigo, B. (1997) *Environmental Science: a global concern.* Fourth Edition. McGraw Hill, New York. p.623.

12 Ballinger, J., Prasad, D. & Rudder, D. (1997) *Energy Efficient Australian Housing.* Second Edition. Australian Government Publishing Service, Canberra, Australia. p.236.

13 Roskill, J. (1971) *Commission on the Third London Airport – Report.* Her Majesty's Stationery Office, London, UK.

14 Alcoa of Australia Ltd., Kinhill Planners Pty. Ltd. (1980) *Alcoa Portland Aluminium Smelter: Environmental Effects Statement and Draft Environmental Impact Statement.* Second Printing. Alcoa of Australia Ltd., Melbourne, Australia.

15 Abelson, P.W. (1979) *Cost Benefit and Environmental Problems.* Saxon House, UK.

16 Langston, C. (1996) The Application of Cost-Benefit Analysis to the Evaluation of Environmentally-Sensitive Projects. Proceedings of CIB meetings held at RMIT 13–16 February 1996.

17 Thomas, I. (1998) *Environmental Impact Assessment in Australia: theory and practice.* The Federation Press, Leichhardt, Australia.

18 Griffith, A. (1994) *Environmental Management in Construction.* The Macmillan Press Ltd, London, UK.

19 Truinius, W. (1999) *Environmental Assessment in Building and Construction: goal and scope definition as key to methodology choices.* Doctoral Thesis, KTH Building Materials Sweden. Taken from ISO 1040.

20 Langston, C. & Ding, G. (2001) *Sustainable Practices in the Built Environment.* Second edition. Butterworth Heinemann, Oxford, UK. Chapter 19.

INDEX

Index